T0311591

ENGINEERING AND TECHNICAL DEVELOPMENT FOR A SUSTAINABLE ENVIRONMENT

ENGINEERING AND TECHNICAL DEVELOPMENT FOR A SUSTAINABLE ENVIRONMENT

Edited by
Dzaraini Kamarun, PhD
Ramlah Mohd. Tajuddin, PhD
Bulan Abdullah, PhD

Apple Academic Press Inc.   Apple Academic Press Inc.
3333 Mistwell Crescent   9 Spinnaker Way
Oakville, ON L6L 0A2 Canada   Waretown, NJ 08758 USA

© 2018 by Apple Academic Press, Inc.

First issued in paperback 2021

Exclusive worldwide distribution by CRC Press, a member of Taylor & Francis Group
No claim to original U.S. Government works

ISBN 13: 978-1-77-463674-9 (pbk)
ISBN 13: 978-1-77-188521-8 (hbk)

Library and Archives Canada Cataloguing in Publication

Engineering and technical development for a sustainable environment/ edited by Dzaraini Kamarun, PhD, Ramlah Mohd. Tajuddin, PhD, Bulan Abdullah, PhD.
Includes bibliographical references and index.
Issued in print and electronic formats.
ISBN 978-1-77188-521-8 (hardcover).--ISBN 978-1-315-20732-2 (PDF)
1. Sustainable engineering. I. Kamarun, Dzaraini, editor II. Tajuddin, Ramlah Mohd, editor III. Abdullah, Bulan, editor
TA170.E53 2018   628   C2017-903062-0   C2017-903063-9

Library of Congress Cataloging-in-Publication Data

Names: Kamarun, Dzaraini, editor. | Tajuddin, Ramlah Mohd, editor. | Abdullah, Bulan, editor.
Title: Engineering and technical development for a sustainable environment /editors, Dzaraini Kamarun, PhD, Ramlah Mohd. Tajuddin, PhD, Bulan Abdullah, PhD.
Description: Oakville, ON, Canada ; Waretown, NJ, USA : Apple Academic Press, [2017] | Includes bibliographical references and index.
Identifiers: LCCN 2017019824 (print) | LCCN 2017023038 (ebook) | ISBN 9781315207322 (ebook) | ISBN 9781771885218 (hardcover : alk. paper)
Subjects: LCSH: Environmental protection. | Technological innovations. | Environmental engineering. | Sustainable engineering. | Manufactures.
Classification: LCC TD170.2 (ebook) | LCC TD170.2 .E53 2017 (print) | DDC 628--dc23
LC record available at https://lccn.loc.gov/2017019824

Apple Academic Press also publishes its books in a variety of electronic formats. Some content that appears in print may not be available in electronic format. For information about Apple Academic Press products, visit our website at **www.appleacademicpress.com** and the CRC Press website at **www.crcpress.com**

CONTENTS

LIST OF CONTRIBUTORS

Bulan Abdullah
Faculty of Mechanical Engineering, Universiti Teknologi MARA, Shah Alam, Malaysia

Ahmad Othman Ahmad
Faculty of Applied Sciences, Universiti Teknologi MARA, Shah Alam, Malaysia

Tengku Nurul Aishah
Malaysian Institute of Transport, Universiti Teknologi MARA, Shah Alam, Selangor, Malaysia

Mohd Fozi Ali
Faculty of Civil Engineering, Universiti Teknologi MARA, Shah Alam, Selangor, Malaysia

Siti Khadijah Alias
Faculty of Mechanical Engineering, Universiti Teknologi MARA, Johor, Malaysia

Sabariah Arbai
Faculty of Civil Engineering, Universiti Teknologi MARA, Shah Alam, Selangor, Malaysia

Azinoor Azida Abu Bakar
Faculty of Civil Engineering, Universiti Teknologi MARA, Shah Alam, Selangor, Malaysia

Mohamad Nor Berhan
Faculty of Mechanical Engineering, Universiti Teknologi MARA, Shah Alam, Malaysia

Rashid Deraman
Faculty of Mechanical Engineering, Universiti Teknologi MARA, Pulau Pinang, Malaysia

Mohammad Fadhil Mohammad
Construction Economics and Procurement Research Group, Centre of Studies for Quantity Surveying, Faculty of Architecture, Planning and Surveying, Universiti Teknologi MARA Shah Alam, Malaysia

Adibatul Husna Fadzil
Faculty of Applied Sciences, Universiti Teknologi MARA, Shah Alam, Malaysia

Z. Faiza
Faculty of Electrical Engineering, Universiti Teknologi MARA, Pulau Pinang, Malaysia

Zulkifli Abdul Ghaffar
Faculty of Mechanical Engineering, Universiti Teknologi MARA, Shah Alam, Malaysia

Ahmad Hussein Abdul Hamid
Faculty of Mechanical Engineering, Universiti Teknologi MARA, Shah Alam, Malaysia

Wan Asma Ibrahim
Forest Research Institute of Malaysia, Kepong, Malaysia

Muhammad Hussain Ismail
Faculty of Mechanical Engineering, Universiti Teknologi MARA, Shah Alam, Malaysia

Muhammad Hafizuddin Jumadin
Faculty of Mechanical Engineering, Universiti Teknologi MARA, Shah Alam, Malaysia

Noor Khadijah Mustafa Kamal
Faculty of Applied Sciences, Universiti Teknologi MARA, Shah Alam, Malaysia

Dzaraini Kamarun
Faculty of Applied Sciences, Universiti Teknologi MARA, Shah Alam, Malaysia

N. A. Kamarzaman
Faculty of Electrical Engineering, Universiti Teknologi MARA, Pulau Pinang, Malaysia

Salmiah Kasolang
Faculty of Mechanical Engineering, Universiti Teknologi MARA, Shah Alam, Malaysia

Karimah Kassim
Faculty of Applied Sciences, Universiti Teknologi MARA, Shah Alam, Malaysia

Khairi Khalid
Faculty of Civil Engineering, Universiti Teknologi MARA, Shah Alam, Selangor, Malaysia

Khalilah Abdul Khalil
Faculty of Applied Sciences, Universiti Teknologi MARA, Shah Alam, Malaysia

Rohana Mahbub
Construction Economics and Procurement Research Group, Centre of Studies for Quantity Surveying, Faculty of Architecture, Planning and Surveying, Universiti Teknologi MARA Shah Alam, Malaysia

Jamaluddin Mahmud
Faculty of Mechanical Engineering, Universiti Teknologi MARA, Shah Alam, Malaysia

N. Mohamad
School of Electrical and Electronic Engineering, Universiti Sains Malaysia, Pulau Pinang, Malaysia

Kamaruzzaman Mohamed
Faculty of Civil Engineering, Universiti Teknologi MARA, Shah Alam, Selangor, Malaysia

Zainab Mohamed
Faculty of Civil Engineering, Universiti Teknologi MARA, Shah Alam, Selangor, Malaysia

Tengku Elida Tengku Zainal Mulok
Faculty of Applied Sciences, Universiti Teknologi MARA, Shah Alam, Malaysia

Muhammad Faiz Musa
Construction Economics and Procurement Research Group, Centre of Studies for Quantity Surveying, Faculty of Architecture, Planning and Surveying, Universiti Teknologi MARA Shah Alam, Malaysia

Noorasyikin Mohammad Noh
Institute for Infrastructure Engineering and Sustainable Management, Universiti Teknologi MARA Shah Alam, Malaysia

Siti Noor Azizzati Mohd Noor
Faculty of Mechanical Engineering, Universiti Teknologi MARA, Shah Alam, Malaysia

N. Othman
Faculty of Electrical Engineering, Universiti Teknologi MARA, Pulau Pinang, Malaysia

Nor Faiza Abd Rahman
Faculty of Civil Engineering, Universiti Teknologi MARA, Shah Alam, Selangor, Malaysia

Kalavathy Ramasamy
Faculty of Civil Engineering, Universiti Teknologi MARA, Shah Alam, Selangor, Malaysia

S. S. Ramli
Faculty of Electrical Engineering, Universiti Teknologi MARA, Pulau Pinang, Malaysia

Siti Nur Ridhwah Muhamed Ramli
Faculty of Applied Sciences, Universiti Teknologi MARA, Shah Alam, Malaysia

Nik Roslan Nik Abdul Rashid
Faculty of Applied Sciences, Universiti Teknologi MARA, Shah Alam, Malaysia

Muhammad Zamir Abd Rasid
Malaysia Agriculture Research and Development Institute (MARDI), Malaysia

Razmi Noh Mohd Razali
Faculty of Mechanical Engineering, Universiti Teknologi MARA, Shah Alam, Malaysia

Muhammad Solahuddeen Mohd Sabri
Faculty of Civil Engineering, Universiti Teknologi MARA, Shah Alam, Selangor, Malaysia

Sabiha Hanim Mohd Salleh
Faculty of Applied Sciences, Universiti Teknologi MARA, Shah Alam, Malaysia

Siti Noriah Mohd Shotor
Faculty of Applied Sciences, Universiti Teknologi MARA, Shah Alam, Malaysia

Mohd Khairi Taib
Faculty of Mechanical Engineering, Universiti Teknologi MARA, Shah Alam, Malaysia

Ramlah Mohd. Tajuddin
Faculty of Civil Engineering, Universiti Teknologi MARA, Shah Alam, Selangor, Malaysia

Nazyra Abdul Wahab
Faculty of Civil Engineering, Universiti Teknologi MARA, Shah Alam, Selangor, Malaysia

Mohd Reeza Yusof
Construction Economics and Procurement Research Group, Centre of Studies for Quantity Surveying, Faculty of Architecture, Planning and Surveying, Universiti Teknologi MARA Shah Alam, Malaysia

Siti Hajar Mohd Yusop
Faculty of Mechanical Engineering, Universiti Teknologi MARA, Shah Alam, Malaysia

Engku Zaharah Engku Zawawi
Faculty of Applied Sciences, Universiti Teknologi MARA, Shah Alam, Malaysia

Fatimah Zuber
Faculty of Applied Sciences, Universiti Teknologi MARA, Shah Alam, Malaysia

LIST OF REVIEWERS

Sunhaji Kiyai Abas
Faculty of Mechanical Engineering, Universiti Teknologi MARA (UiTM), 40500, Shah Alam,
E-mail: sunhajiabas@yahoo.com

Nik Rosli Abdullah
Faculty of Mechanical Engineering, Universiti Teknologi MARA (UiTM), 40500, Shah Alam,
E-mail: nikrosli@salam.uitm.edu.my

Rohaya Ahmad
School of Chemistry and Environment, Faculty of Applied Sciences, UiTM Shah Alam, Malaysia,
E-mail: rohayaahmad@salam.UiTM.edu.my

Yasmin Ashaari
Faculty of Civil Engineering, Universiti Teknologi MARA, 40450 Shah Alam, Malaysia

Haryari Awang
Faculty of Civil Engineering, Universiti Teknologi MARA, 40450 Shah Alam, Malaysia

Abdul Ghalib @ Tham Hock Khan
Faculty of Mechanical Engineering, Universiti Teknologi MARA (UiTM), 40500, Shah Alam,
E-mail: gtkhan51@gmail.com

Dato' Roslan Hashim
Professor of Geotechnical Engineering, Department of Civil Engineering, University of Malaya,
E-mail: roslan@um.edu.my

Azianti Ismail
Faculty of Mechanical Engineering, Universiti Teknologi MARA (UiTM), Johor (Pasir Gudang),
E-mail: azianti106@johor.uitm.edu.my

Muhammad Hussain Ismail
Faculty of Mechanical Engineering, Universiti Teknologi MARA (UiTM), 40500, Shah Alam,
E-mail: hussain305@salam.uitm.edu.my

Wirachman Wisnoe
Faculty of Mechanical Engineering, Universiti Teknologi MARA (UiTM), 40500, Shah Alam,
E-mail: wira_wisnoe@yahoo.com

Noriah Yusoff
Faculty of Mechanical Engineering, Universiti Teknologi MARA (UiTM), 40500, Shah Alam,
E-mail: noriahyusoff@salam.uitm.edu.my

Anizah Kalam
Faculty of Mechanical Engineering, Universiti Teknologi MARA (UiTM), 40500, Shah Alam,
E-mail: anizahkalam@salam.uitm.edu.my

Lee Wei Koon
Faculty of Civil Engineering, Universiti Teknolohi MARA, 40450 Shah Alam, Malaysia

Ismail Rahmat
Fakulti Senibina, Perancangan and Ukur, Kompleks Tahir Majid, Universiti Teknologi MARA, 40450 Shah Alam, Selangor Darul Ehsan, E-mail: ismail046@salam.UiTM.edu.my

Azli Abd Razak
Faculty of Mechanical Engineering, Universiti Teknologi MARA (UiTM), 40500, Shah Alam, E-mail: azlirazak@salam.uitm.edu.my

Juri Saedon
Faculty of Mechanical Engineering, Universiti Teknologi MARA (UiTM), 40500, Shah Alam, E-mail: jurisaedon41@salam.uitm.edu.my

Sabu Thomas
Centre for Nanoscience and Nanotechnology, School of Chemical Sciences, Mahatma Gandhi University, Priyadarshini Hills P.O., Kottayam, Kerala, India – 686560, E-mail: sabuchathukulam@yahoo.co.uk

Nor' Aini Wahab
Faculty of Mechanical Engineering, Universiti Teknologi MARA (UiTM), 40500, Shah Alam, E-mail: noraini416@salam.uitm.edu.my

LIST OF ABBREVIATIONS

AIAA	American Institute of Aeronautics and Astronautics
AR	as-received
ATR	attenuated total reflectance
BEEC	Building Energy Efficiency Certificate
BIOECODS	bio-ecological drainage system
BSA	bovine serum albumin
BuA	butyl acrylate
CB	carbon black
CBD	Commercial Building Disclosure
CO_2	carbon dioxide
COAG	Council of Australian Government
CRI	cure rate index
DID	Department of Irrigation and Drainage
DMAC	dimethyl acetamide
DMAc	N,N-dimethylacetamide
DSC	differential scanning calorimetry
EHA	2-ethyl hexyl acrylate
EMS	environmental management system
EPDM	ethylene propylene diene monomer
EPDM-r	ethylene propylene diene terpolymer residues
ERGS	exploratory research grant scheme
FEF	fast extruding furnace
FESEM	field emission scanning electron microscope
FGC	filter glass crucible
FL	fusion line
FTIR	Fourier transform infrared
GBI	green building index
GCC	ground calcium carbonate
GHG	green house gas
GMAW	gas metal arc welding

H_2O	water vapor
HEC-HMS	Hydrologic Engineering Centre-The Hydrologic Modeling System
IBS	industrialized building system
ICRCL	Inter-department Committee on the Redevelopment of Contaminated Land
IEQ	indoor environmental quality
IMMB	Institute Molecular Medical Biotechnology
IRCDIP	International Research Centre for Disaster Prevention
IWCS	Info Works Collection System
LCA	life cycle assessment
LPR	linear polarization resistance
MBI	Modular Building Institute
MBRs	membrane bioreactors
MMD	Meteorological Department
MOE	Ministry of Education
MOHE	Ministry of Higher Education
MSM	minimal salt medium
MSMA	Storm Water Management Manual for Malaysia
NABERS	National Australian Built Environment Rating System
NKVE	New Klang Valley Expressway
NMP	N-methyl–2-pyrrolidone
NMR	nuclear magnetic resonance
POME	palm oil mill effluent
PSF	polysulfone
PVP	polyvinyl pyrrolidone
RBF	round bottom flask
REDHA	Real Estate and Housing Developers Association Malaysia
RMI	Research Management Institute
SCE	saturated calomel electrode
SCS-CN	Soil Conservation Service Curve Number Method
SEM	scanning electron microscopy
SHT	solution heat-treated
SRHT	stress relief heat treatment
SWAT	Soil Water Assessment Tools

SWMM	Stormwater Management Model
TG	thermogravimetric
TGA	thermogravimetric analyzer
UF	ultrafiltration
UiTM	Universiti Teknologi Mara
USLE	Universal Soil Loss Equation
UTS	ultimate tensile strength
XRD	x-ray diffraction

PREFACE

Advancement in engineering and technology encompasses both high technological innovation and creations of modern industry and alternative technologies dominant in developing and developed countries.

Alternative technological advancement can be intermediate in nature and indigenous and cost-effective, and they are most often driven by dominant societal interest and geographical status. When used in synergy with high-tech developments, alternative technology could bring about changes that could be steered toward human needs and sustainable environment.

This book, *Engineering and Technical Development for a Sustainable Environment*, discusses several issues of alternative technologies and its developments in several areas of science and engineering with particular attention to the utilization of bioresources for the achievement of a sustainable environment. Several engineering and technological processes are included to mark their importance in product performance and preservation of the environment.

Some of the topics covered in this book are: (i) strategies for the management of rain and ground water for consumption; (ii) surface water retention for urban storm water management; (iii) scientific and engineering approaches on the prevention of flood and landslides in the tropics; (iv) wind power generation; (v) soil evaluation of contamination due to heavy metals; (vi) green and sustainable building approaches; (vii) bioethanol production; and (viii) energy conservation techniques.

The book presents several selected case studies of engineering and technological processes in industries that relate to the utilization of bioresources and green engineering practices for higher product performance and efficient processes. Specific examples are reverse engineering for product reformulation, bioresources for biofuel production, novel cyclic thiourea for corrosion inhibition, and green mechanical processes and techniques.

Refreshing and informative, *Engineering and Technical Development for a Sustainable Environment* revisits conventional approaches of managing natural agents (such as wind, rain and groundwater resources as well as wastewater treatment) in light of current sustainable-oriented techniques using modern scientific concepts and strategies. It presents in-depth evaluations and analyses using systematic up-to-date scientific and engineering tools.

ABOUT THE EDITORS

Dzaraini Kamarun, PhD

Associate Professor cum Lecturer, Universiti Teknologi MARA, Malaysia

Dzaraini Kamarun, PhD, is an Associate Professor cum Lecturer at the Universiti Teknologi MARA, Malaysia. She has experience in writing and editing academic literature for tertiary education and was the chief editor for her first published book, *Progress in Polymer and Rubber Technology.* She has published many research articles in journals and proceedings as well as in several local magazines. She has over 30 years of teaching and writing experience. She received her PhD in Biosensors from the Queen Mary University of London, UK.

Ramlah Mohd. Tajuddin, PhD

Associate Professor cum Lecturer at Universiti Teknologi MARA, Malaysia

Ramlah Mohd. Tajuddin, PhD, currently works as an Associate Professor cum Lecturer at the Universiti Teknologi MARA, Malaysia. She has experience in writing and editing academic literature for tertiary education and is the author of the book *Introduction to Sustainable Campus Initiatives.* She has also published many research articles in journals and proceedings. She has 28 years of teaching experience as well as writing and editing experience. She received her PhD in Civil Engineering-Membrane Technology for Water and Wastewater Treatment from the Universiti Technologi Malaysia, Malaysia.

Bulan Abdullah, PhD

Senior Lecturer and Coordinator for Postgraduate Programme, Faculty of Mechanical Engineering, Universiti Teknologi MARA, Malaysia

Bulan Abdullah, PhD, is a Senior Lecturer and Coordinator for the Postgraduate Programme in the Faculty of Mechanical Engineering at the

Universiti Teknologi MARA in Malaysia. Her research experience is in advanced material characterization, advanced manufacturing, total quality management, and statistical quality control. She currently teaches courses in manufacturing processes, total quality management, research methodology, and industrial projects.

CHAPTER 1

POTENTIAL OF DIFFERENT GRASS SPECIES ON SURFACE WATER RETENTION TO IMPROVE URBAN STORM WATER MANAGEMENT

RAMLAH MOHD. TAJUDDIN,[1] TENGKU NURUL AISHAH,[2] and NAZYRA ABDUL WAHAB[1]

[1]*Faculty of Civil Engineering, Universiti Teknologi MARA, Shah Alam, Selangor, Malaysia*

[2]*Malaysian Institute of Transport, Universiti Teknologi MARA, Shah Alam, Selangor, Malaysia*

CONTENTS

OVERVIEW

An increasingly industrialized global economy over the last century has led to a dramatic increment of the impervious area which contributes to the rise of volume and velocity of urban storm water. It results in the rapid discharges into stream and river that caused flash flood events if there is no provision made for collection, storage and proper discharge. Vegetation can help in enhancing the infiltration and surface roughness and reduces the kinetic impact of raindrops thus enhancing the groundwater recharge and reducing flash flood. In addition, the vegetation can also intercept the raindrops and stored on its outer surface for certain period of time. This will further reduced the rate and volume of the stormwater runoffs. Considering the large area covered by these vegetations, a significant effect can be obtained through interception storage on flash flood resulted from rainfall events. However the effects of different characteristics of plants cover on the rate and volume of surface runoff were very limited. Therefore this study focuses on the interception storage performance of local grass species which normally used in Malaysia. Three different types of grass species; Pearl grass *(Axonopuscompressus)* (dwarf), Philippine grass *(Zoysiamatrella)* and Cow grass *(Axonopuscompressus)* were chosen in order to determine their potential in reducing the runoffs. An artificial bed slope were fabricated in the hydrology laboratory of Universiti Teknologi MARA, Shah Alam, Selangor, Malaysia to simulate the rainfall event at site but having control on the rainfall intensity and the plants. Two different rainfall intensities of 17 mm/hr and 166 mm/hr were applied. The grasses were planted in a bed slope with an artificial precipitation. The volume of rainfall was measured as the system is running to determine the water retention capacity. Results showed that rainfall intensity did give significant differences to the percentage reduction of surface runoffs. Laboratory experimental results showed that Philippine grass *(Zoysiamatrella)* which has the highest dense patch which showed the highest potential of reducing the volume of surface runoff.

1.1   INTRODUCTION

An increasingly industrialized global economy over the last century has led to dramatically increment of the impervious area which contributes to the rise of volume and velocity of urban storm water. This is due to the increased in the construction of the paved area which lead to a great significant changes on the hydrologic and hydraulic characteristics of the catchment. Therefore, the urban storm water management is very important, particularly for developing countries in order to prevent the event of flashflood events, soil erosion, water pollution and eutrophication of water bodies that are resulted by urbanization.

As in Malaysia, the population of urban dwellers shows a progressive increment. In 1970 the population was 26.8% and it increases to 35.8% in 1980. The population continuously increased to 50.7% in 1991. It is predicted to exceed 65% by the year 2020 [1]. The continuous growth of urban areas significantly contributes to the increment of the impervious area and hence resulted in the frequent occurrence of flash flood. This is proven by a study in Belo Horizonte [2] which shows the correlation between population growth and flood events. Numerous studies in Australia have revealed that only 10% of rainfall runoffs were recorded from areas covered with natural bushes and trees in contrary with urban areas, the rainfall runoffs from this area can reached up to 90% due to the impervious surfaces in urban areas [3, 4].

Flood is a major threat to Malaysia. Flood events have becoming more frequent as rapid as urbanization rate took place. Flash flood in February 2006 that occurred in the Shah Alam municipality has caused damages to properties worth more than 100 million ringgit where 10,000 citizens were involved and the highways links were interrupted as the same goes to the commuter railway. Flood in year 2007 in Terengganu entangled 1,147 victims in Terengganu and 1,121 victims in Sandakan, Sabah in the year 2010 [5].

Runoff or urban storm water is generated by rainfall, which cannot be infiltrated into the ground when it falls onto impervious surface area where there is no provision is made for its drainage. The increased of paved area, hence caused a tremendous increased in surface runoff which caused flash flood. Proper understanding of nature of the land surface to receive the rainfall can help to improve the management of stormwater which can help to reduce the occurrence and volume of flood.

According to Yu [6] the vagaries of climate, geology, topography, degree of imperviousness such as roofs, pavements and roads and human activities, including urban management has a significant effect to the quantity and quality of urban storm water. Due to the increment in runoff volumes and pollutant discharges, the need of planning, design and management of storm water becoming more crucial [1]. Therefore, the Department of Irrigation and Drainage (DID) Malaysia has come up with the new Storm Water Management Manual for Malaysia (MSMA) which draws on various approaches of "Best Management Practices" in managing the storm water. This approach is more environmental friendly. An example of an integrated approach of managing storm water has been applied in the previous study by DID Malaysia using Bio-Ecological Drainage System (BIOECODS) where the use of swale and cow grass is being planted in the drainage system in order to control the quantity and quality of storm water.

The use of vegetation or green technology can be said as a sustainable urban storm water approach. By planting ground covers, it will enhance in controlling erosion and runoff by protecting the soil from raindrop impact, reducing soil compaction and improving infiltration. In terms of esthetic value, vegetation can give attractive scenery with a variety of color, texture and height and economical compared to hard engineering [7]. Vegetation plays an important role in decreasing soil detachment and transport from project sites where the soil surface has been disturbed by human activities. Vegetation promotes long-term protection of the soil surface by providing leaf cover that intercept precipitation and by establishing roots, which aid soil structure development, thereby increasing infiltration and soil stability. Vegetation also provides a viable alternative to much synthetic means of erosion control, increases biodiversity, and increases the esthetic value of project landscapes [8, 9].

Planting ground covers on slopes or bare areas helps control erosion and runoff because plant roots hold the soil in place and the leaves protect the soil from the impact of raindrops, reducing soil compaction and improving the speed with which water soaks into the ground. Ground covers can produce attractive patterns with variation in height, texture and color. They also conserve soil moisture, reduce maintenance in narrow odd-shaped areas where mowing, edging and watering might be difficult; reduce heat, glare, noise, and dust; and block foot traffic without blocking the view [7]. However, plant species likely vary in their performance relative to these functional goals [10, 11]. A study by Gerrits et al. [12] proves that the storage capacity differs with vegetation type and season. A study by Andersen et al. [13] has proven the potential of vegetation as green roof to manage storm water. In this study, moss species can hold high capacities of water by storing 8–10 times of their weight in water compared to only 1.3 times for typical green roof medium. Mock-up roof sections composed of mosses, and medium had delayed and reduced runoff flows. Many research works had looked into plant influence on erosion control [14, 15].

1.2   MATERIALS AND METHODS

The runoff collection system used in this study is made from plywood and was fabricated in the Hydrology Laboratory, Faculty of Civil Engineering, UiTM. The fabrication of the system according to the lab scale runoff collection system which consists of four beds slope in which it is 2.0 m length × 0.6 m width × 0.15 m depth each. Each area of the bed slope will represent the catchment area for the test. The slope used is at 1V:2H [8, 16] whereby gives to the degree of 27°.

The rainfall simulator devices were self-contained units that consist of sprinkler, pump, piping, and sump. The sprinklers were placed at least 0.6 m above the soil surface to create an impact velocity nearly equal to that of a natural raindrop as shown in Figure 1.1. Figure 1.2 shows the plan view of the runoff collection system which consist of four boxes

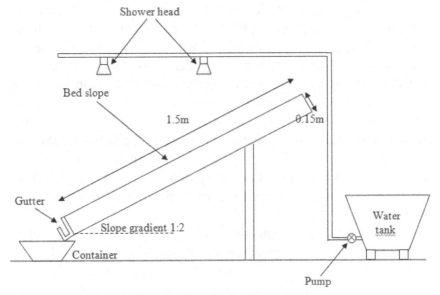

FIGURE 1.1   Schematic diagram of runoff collection system.

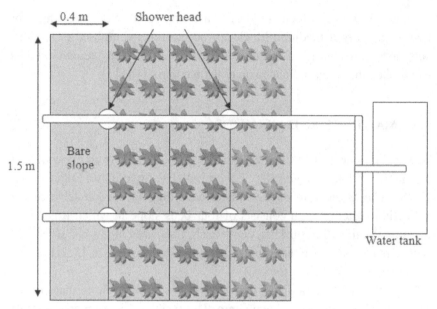

FIGURE 1.2 Plan view of runoff collection system.

which were equipped by a control box (bare slope), and the other three are planted with three different types of grass species selected.

Three types of grass species were selected and planted in the Runoff Collection System as shown in Figures 1.1 and 1.2. The water retention capacity of the plants was measured when it was subjected to different simulated rainfall intensity. A layer of aggregates about 25.4 mm was placed on the bed slope followed by top soil and being spread on the test bed surface to a uniform depth of approximately 15 cm to allow plant growth. On the bare soil plots, the compacted soil was covered with a layer of transparent plastic to act as an impervious surface area. The soil type selected for use in the indoor laboratory test program was top soil typical of fill material used for planting purposes. The soil sample's characterization was identified through particle size distribution analysis in accordance with BS 1377 Part 2 1990.

After laying the soil in the box, the box was raised from a horizontal position to the 1:2 slope gradients as shown in Figure 1.1. One of the bed slopes will be used for impervious surface, and the other three were planted with selected grass species. Three different grass species selected were Pearl grass (*Axonopuscompressus dwarf*)), Philippine grass (*Zoysiamatrella*) and Cow grass (*Axonopuscompressus)* as shown in Figures 1.3(a), (b) and (c).

These grasses were showered by simulated rainfall at intensity of 17 mm/hr and 166 mm/hr. These two different rainfall intensities are to

FIGURE 1.3 (a) Pearl grass.

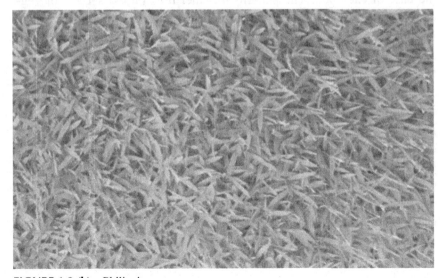

FIGURE 1.3 (b) Philippine grass.

FIGURE 1.3 (c) Cow grass.

represent low and high rainfall intensity. Runoffs were then collected from the base of each box. The patterns of water flow on the surface of vegetative cover were observed and their flow pattern were compared. The flow pattern for each plant, according to its physical characteristics such as roots types and width of the leaf. The process were repeated for several times at 5, 10, 15, 20 and 30 minutes.

Runoff and sediment samples from each of the boxes were collected in separate containers positioned at the lower end of the test bed as shown in Figure 1.1. The total water that can retain by plants was calculated by comparing the total volume of runoff from each treatment to the control box collected. The output channel was constructed at the surface level so that the water that infiltrate into soil did not disrupt the total volume of surface runoff collected. The physical characteristics of each grass species were being observed and recorded according to formation of leaves, the sizes of leaves and the roots. Then, the relationships between volumes of runoff reduction with physical characteristics of plants were determined.

1.3   RESULTS ANALYSIS AND DISCUSSIONS

Runoff patterns collected from the Runoff Collection Systems when subjected to the two intensities of rainfall events were shown in Figures 1.4 and 1.5. These results show that all grass types are able to retain more rainfall when they were exposed to low intensity of rainfall. Cow grass showed a constant reduction of runoff for all the five readings taken in the low intensity event where the percentage reductions range between 93% and 99%. It is followed by Philippine grass which showed a percentage runoff reduction about 80% to 99% in range. While the Pearl grass showed the least potential, but still can be said as one of the best vegetative cover that can be used because of the percentage runoff reduction is range from 40% to 90% which it can help to reduce half of the total runoff. These make cow grass as the best vegetative cover at low simulated rainfall intensity event compared to Pearl grass and Philippine grass. The dense patch of cow grass's canopy and broader leaves are believed to be the significant to the enormous reduction of surface runoff. It helps to retain a lot amount of surface runoff by trapping the water between the leaves. As for the Philippine grass, although

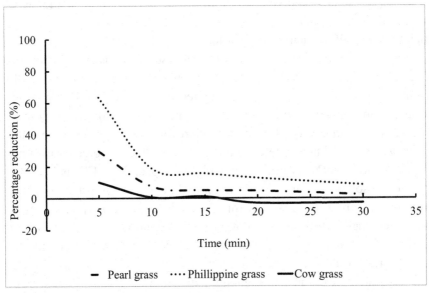

FIGURE 1.4   Percentage of runoff reduction at low simulated (17 mm/hr) rainfall intensity (%).

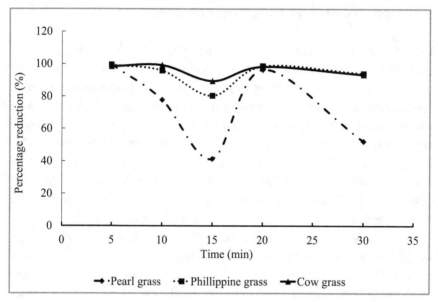

FIGURE 1.5   Percentage of runoff reduction at high simulated (166 mm/hr) rainfall intensity (%).

it leaves has smaller and vertically formed like needles its' fine and dense root could help to retain more water compared to Pearl grass due to allowing the water to infiltrate into the soil faster.

On the other hand, Pearl grass which has short, rounded and horizontally shape type of leaves caused it to form a flat surface that eased the rainfall flow onto pearl grass surface rather than infiltrate into the ground. Contrary in the high-simulated rainfall intensity event, cow grass showed the least potential vegetative cover where the results were almost the same as in control box. There were negative readings recorded which are −3.07% and −2.69% for minute 20 and 30, respectively. This is believed due to the soil that saturated with water and the excessive water collected was coming from the saturated soil. As for the other two types of grass, the percentage reduction is getting lower throughout the time taken as it showed that there is some limitation the plants can help in retaining the surface runoff. Overall, Philippine grass can be said as the most potential vegetative cover in this high-simulated rainfall intensity event with a percentage runoff reduction range of 8% to 63%.

TABLE 1.1   Observation of Physical Characteristics of Pearl Grass (*Axonopuscompressus* – *dwarf*), Philippine Grass (*Zoysiamatrella*) and Cow Grass (*Axonopuscompressus*)

Plant name	Leaves	Roots
Axonopuscompressus (dwarf) (Pearl grass)	Dense patch leaves are in round shape.	Rooting firmly, and trailing horizontally and splitting at each end.
Zoysiamatrella. (Philippine grass)	Leaves are smaller and vertically form like needles and harder texture.	Dense and fine roots
Axonopuscompressus (cow grass)	Dense patch leaves are wide and long shape compared to pearl grass.	Rooting firmly, and trailing horizontally and splitting at each end.

Table 1.1 displayed the comparisons made based on observation on the different physical characteristics of grass species as shown in Figures 1.3(a), 3(b) and 3(c). These figures showed that there was a significant difference in the amount of water runoff between the three vegetative covers.

Formation of leaf canopy either vertically or horizontally, dense patch, the size of the leaf and dense root should be considered as to choose the best vegetative cover in the urban area. This shows the structural qualities of plants have a great significance to the volume of water retained. It is well recommended that grasses can be one of the best potential vegetative cover for effective storm water management, especially in an urban area that has a limited area to plant a bigger vegetative cover. In future research, water runoff experiments should be conducted under a real storm events and natural rain and over the short and long term so that the plants used can reach their maximum potential. It would also be beneficial to study other local plants potential for the urban storm water management.

1.4   CONCLUSION

Physical characteristics of plant play an important role in reducing the surface runoff volume during low or heavy rainfall. The leaf structure, texture and the stem height from the ground surface are the three dominant factors influencing the interception storage quantity and hence the surface runoff volume. The interception storage capacity and resident time can play an important role to reduce the quantity and flow rate of surface runoff in order to manage flash flood events in urban areas where most of the surface had been converted to impervious surface. However, it can be seen that each plant has its maximum interception storage capacity. During heavy rainfall events, the reduction of percentage storage reduces. This is because the plants have reached its maximum storage ability. Therefore, it is advisable to have varieties of plant species to be used as vegetative cover to manage storm surface runoff since each plant has different potential interception storage capacity.

ACKNOWLEDGMENT

Special thanks to Mr. Noorul Hilmi and Mr. Zarinnizar for assistance with the runoff collection design. The financial support of the Research Management Institute (RMI) of Universiti Teknologi Mara (UiTM), Malaysia, in carrying out this study is gratefully acknowledged.

KEYWORDS

- cow grass
- flash flood
- interception
- pearl grass
- Philippine grass
- storm water management

REFERENCES

1. Zakaria, N. A., Ab. Ghani, A., Abdullah, R., Sidek, L. M., Kassim, A. H., & Ainan, A. (2004). MSMA: a new urban stormwater management manual for Malaysia. *Advances in Hydroscience, Vol. 4.*

2. Baptista, M. B., Nascimento, N. O., Ramos, M. H. D., & Champs, J. R. B. Aspectos da evolução da Urbanização e dos problemas de inundaçõesem Belo Horizonte (Aspects of urban development and floods in Belo Horizonte). In: Braga, C. Tucci, & M. Tozzi, (ed.). *Drenagem Urbana- Gerenciamento, Simulação e Controle*, 39–50. ABRH, Porto Alegre, Brazil, 1998.

3. Environment Australia. *Introduction to Urban Stormwater Management in Australia (Canberra: Environment Australia)*, 2002.

4. Laurenson, E. M., Codner, G. P., & Mein, R. G. (2005). *Urban Storm Water in Sydney: Present Infrastructure and Management Practices and Trends in Stormwater.*

5. Wardah, T., Abu Bakar, S. H., Bardossy, A., & Maznorizan, M. (2008). Use of geostationary meteorological satellite images in convective rain estimation for flash-flood forecasting. *Journal of Hydrology 356*, 283–298.

6. Yu, X. (2008). Use of low quality water: an integrated approach to urban stormwater management (USM) in the greater metropolitan region of Sydney (GMRS). *Journal of Environmental Studies, 65(1)*, 119–137.

7. Relf, D. (2001). Reducing Erosion and Runoff. Environmental Horticulture. *Virginia Tech Publication Number 426–722.*

8. Hallock, B. G., Dettman, K., Rein, S., Curto, M., & Scharff, M. (2003). Effectiveness of native vegetation planting techniques to minimize erosion. *American Water Resources Association, 2003 Annual Water Conference*, Nov 2–5, San Diego, California.

9. Morgan, R. P. C. (1995). *Soil Erosion and Conservation*. Longman Group Limited, London.

10. Dunnet, N., Nagase, A., Booth, R., & Grime, P. (2008). Influence of vegetation composition on runoff in two simulated green roof experiments. *Urban Ecosystem, 11*, 385–398.

11. Monterusso, M. A., Rowe, D. B., Rugh, C. L., & Russel, D. K. (2004). Runoff water quantity and quality from green roof systems. *ActaHort (ISHS), 639*, 369–376.

12. Gerrits, A. M. J., Pfister, L., & Savenije, H. H. (2010). *Spatial and Temporal Variability of Canopy and Forest Floor Interception in a Beech Forest*. Diss. Abstr. 1–540.

13. Anderson, M., Lambrinos, J., & Schroll, E. (2010). The potential value of mosses for storm water management in urban environments. *Urban Ecosystem, 13*, 319–332.

14. Truong, P., & Loch, R. (2004). *Vetiver System for Erosion and Sediment Control*. Proc. 13th Intl. Soil Conserv. Org. Conf., Brisbane, Qld., Australia.

15. Xu, Liyu. (2003). *Vetiver System for Agriculture Production*. Proc. ICV-3, Guangzhou, China.

16. *Urban Stormwater Management Manual for Malaysia*. Department of Irrigation and Drainage, 2001.

CHAPTER 2

INFILTRATION APPROACH IN HYDROLOGICAL MODELING FOR FLOOD AND EROSION PREVENTION

KHAIRI KHALID,[1] MOHD FOZI ALI,[2] NOR FAIZA ABD RAHMAN,[2] and MUHAMMAD ZAMIR ABD RASID[3]

[1]*Faculty of Civil Engineering, Universiti Teknologi MARA Pahang, Malaysia*

[2]*Faculty of Civil Engineering Universiti Teknologi MARA, Shah Alam, Malaysia*

[3]*Malaysia Agriculture Research and Development Institute (MARDI), Malaysia*

CONTENTS

OVERVIEW

Infiltration is one of the major components of the hydrologic cycles. In watershed management, prediction of flooding, erosion, and pollutant transport all depend on the rate of runoff that is directly affected by the rate of infiltration. By understanding how infiltration rates are affected by surface condition, the measure can be taken to increase infiltration rates and reduce the erosion and flooding caused via overland flow. The paper updates several different of infiltration approaches in a few hydrological models for Malaysian climate. It is found that most of the hydrologic model performance was conducted using several statistical tests and no further study to relate the infiltration processes towards the performance of the generated surface runoff. The paper will benefit to the soil and conservation engineer by providing the SCS-CN infiltration approach in approximate physical based hydrological model to be used for the river catchment in Malaysia and each antecedent water condition.

2.1   INTRODUCTION

Flooding is one of the most hazardous natural events, and it is often responsible for loss of life and severe threat to infrastructures and environment. Activities in the floodplain and catchment such as land clearing for developments may increase the magnitude of the flood. In the past, nature took care of itself as vast expanses of forests and wetlands soaked up rainfall excess and delayed the flow of water into the river basin. Absolute control over floods is rarely feasible either physically or economically. On the other hand, flood mitigation measures are undertaken to reduce flood damage to a minimum, consistent with the cost involved. Besides the construction of dams and reservoirs and the improvement of river systems, the processes to increase infiltration and to store the excess water in small ponds and retention basins are being promoted [1]. Land use control, floodplain management, development control and improvement of flood forecasting and warnings are the non-structural measures that need to be highlighted in reducing the flood disaster. Understanding how the river basin reacts to describe the flood hydrograph becomes a crucial part before any project of flood mitigation approach is implemented.

The physiographic of the river basin and climatic factors play the most critical factors affecting the flood hydrograph of the river basin. These include the basin characteristics, infiltration characteristics and channel characteristics.

There are numbers of hydrological models have been used in the water resources study. Liew & Selamat [2] reported that in Malaysia, the Stormwater Management Model (SWMM) and Info Works Collection System (IWCS) are among the most widely used software to model drainage systems. From the literature, it is found that many other hydrological models are also beenutilized for the watershed modeling study in this country. These included a Hydrologic Engineering Centre-The Hydrologic Modeling System (HEC-HMS) software [3–7], followed by the MIKE SHE @ System Hydrologique European (SHE) model [8] and finally the most current model is the Soil Water Assessment Tools (SWAT) software [9, 10]. Table 2.1 shows the model functionality comparison among the most common hydrological models with the variety of the infiltration approaches.

The predominant hydrologic processes include rainfall, interception, evapo-transpiration, infiltration, surface runoff, percolation and subsurface flow. The ability to quantify infiltration is of great importance in the watershed management mostly in the prediction of flooding, erosion,

TABLE 2.1 Model Functionality Comparison

Model Features	HEC HMS	SWMM	IWCS	SWAT	MIKE SHE
Model Type	Lumped/ Distributed	Distributed	Distributed	Distributed	Distributed
Infiltration	SCS-CN	Green-Ampt model	SCS-CN	SCS-CN	Richards Equation
	Initial and uniform loss	Initial loss towards continuing loss model	Green-Ampt model	Green-Ampt model	
	Exponential loss rate		Fixed percentage runoff model		
	Holtan loss rate				
	Green-Ampt loss rate				

and pollutant transport. This process is also necessary for determining the availability of water for the crop growth and to estimate the amount of additional water needed for irrigation. By understanding how infiltration rates are affected by surface conditions, the measure can be taken to increase infiltration rates and reduce the erosion and flooding caused by the overland flow.

Although it is great importance, many hydrologic models still lack proper quantification of infiltration. The widely used hydrologic models use the Soil Conservation Service Curve Number Method (SCS-CN), an empirical formula for predicting runoff from daily rainfall. There are limitations of SCS-CN as determined by the previous researchers. Firstly, it is found that SCS-CN does not reproduce measured runoff from specific storm rainfall events because the time distribution is not considered [11, 12]. The CN method include lack of explicit account for the effect of the antecedent moisture conditions in runoff computation, difficulties in separating storm runoff from the total discharge hydrograph, and runoff processes not considered by the empirical formula [11, 12]. Finally, estimation of runoff and infiltration derived from the CN method may not be representative of the observed value. Croley and He [13] claimed that the use of the Curve Number method may also result in the inaccurate estimation of non-point source pollution rates.

This paper attempts to provide a framework for infiltration issues in the hydrological modeling. The first section gives some basic classification of the infiltration model approaches. This is important for the researcher to embed the infiltration data in the range of hydrological modeling. The second section discusses the factors that contribute to infiltration rate in the catchment. The third section shows hydrological simulation at the upper part of Langat River Basin using SWAT model and a few infiltration issues in the study are discussed. The paper concludes by identifying key issues and gives some directions for future research.

2.2   INFILTRATION MODELING APPROACHES CATEGORY

Infiltration modeling approaches are often separated into three categories; physical based, approximate physical based and the empirical models. The physical based model require solution of the a Richards' equation, which describe water flow in soils in terms of hydraulic conductivity and the soil water pressure as functions of soil water content, for specified boundary conditions. Rawls et al. [14] has claimed that solving this equation is extremely difficult for many flow problems requiring detailed data input and use of the numerical method. Although numerical methods allow the hydrologist to quantify the vertical percolation of water but it is critical for assessment of groundwater recharge [15]. A large quantity and the complexity of the measurements necessary to obtain much of the soil property data required for these numerical solutions impose a more severe limitation that has diminished with time. Consequently, for many applications, equations that simplify the concepts involved in the infiltration process are advantageous.

It has been noted that different approximate equations for infiltration results in different predictions for infiltration rate, time of ponding and time of runoff even when measurements from the same soil samples are used to derive parameter value. There are many factors that contribute to the infiltration rate including time from the onset of rain or irrigation, the initial water content of the soil, hydraulic conductivity, surface conditions, and profile depth, and layering. All the infiltration equations make use of some of these factors in charactering infiltration. A details characteristic of the infiltration approaches is summarized in Table 2.2.

TABLE 2.2   Infiltration Approaches in Hydrological Modeling

Type of Approach	Model	Characteristics
Physically Based Model	1. Richards' equations	Rely more heavily on the soil hydraulic and physical properties occurring within the profile such as saturated hydraulic conductivity, soil moisture gradients, and suction at the wetting front.
Approximate Physical Based	1. Green-Ampt equation 2. Philip and Smith 3. Parlange equation	Models in this category used empirical models with parameters determined in a similar manner. Not requiring the measured infiltration data but based on assumptions that can never be entirely valid.
	4. SCS-CN	Assumption of the homogeneous soils, uniform initial water content and neglect the effect of entrapped air.

TABLE 2.2   (Continued)

Type of Approach	Model	Characteristics
Empirical Models	1. Kostiakov 2. Horton 3. Holtan 4. Phi Index	Rely more on parameters that are determined by curve fitting or estimated by other means and thus may better reflect the effect of differences in surface conditions than the physical model.
		Provide infiltration rates based on measured field data, therefore, provide more realistic estimates but less versatile.

Empirical model tend to be less restricted by the assumptions of soil surface and soil profile conditions, but more restricted by the conditions for which they are calibrated, since their parameters are determined based on the actual field-measured infiltration data. Equations that are physically based approximations use parameters that can be obtained from soil water properties and do not require measured infiltration data. From the type of approaches as discussed above, Table 2.3 highlighted the advantage and disadvantage of the related equations.

TABLE 2.3   Advantage and Disadvantage of the Infiltration Equation

Equation	Advantage	Disadvantage
Richards Equation	1. Uses "measurable" soil and fluid characteristics	1. The nonlinear equation, difficult numerical solution.
	2. Redistributes moisture in the soil profile	2. Difficult to account for spatial and temporal parameter variability.
	3. Details information about moisture profile for the simulation depth	3. Does not directly account macropores.
	4. Can conceptually account for layer soils 2-D or 3-D flow	4. Impact of roots and vegetative characteristics difficult to consider.
Green-Ampt Equation	1. Uses "measurable" soil and fluid characteristics	1. No redistribution of moisture after rainfall.
	2. Simple and easy to solve	2. Difficult to account for spatial and temporal parameter variability.
	3. Closely approximate values obtained using Richards equation	3. Difficult to account for macropores, roots, and vegetative characteristics.

TABLE 2.3   (Continued)

Equation	Advantage	Disadvantage
Holtan's Method	1. Simple 2. Parameter reflects non-ideal conditions 3. Well-suited daily meteorological variables 4. Few parameters required for calibration	1. Assumes moisture is redistributed throughout complete profile. 2. Sensitive to value of D, different values for ET. 3. Need to estimate percolation indirectly.
SCS-CN Method	1. Accounts for all abstractions in single technique 2. Parameters can be estimated from large data set 3. Widely used design procedure for long period 4. Simple and easy to use	1. Crude approach for accounting for soil moisture content. 2. Not as easy to incorporate into continuous simulation models. 3. Does not account for percolation directly.

2.3   FACTOR THAT CONTRIBUTE TO INFILTRATION RATE

Infiltration is the process by which water on the ground surface enters the soil from rainfall, snowmelt or irrigation, from the soil surface into the top layer of the soil. The water then moves from point to point within the soil, and the process is known as a redistribution process. Both of the processes cannot be separated because the rate of infiltration is strongly influenced by the rate of water movement within the soil below. Same factors that control the infiltration rate also have an important role in the redistribution of water below the soil surface during and after infiltration. Skaggs and Khaleel [15] emphasize that an understanding of infiltration and the factors that affect the process is critical in the determination of the surface runoff, understanding the subsurface movement and storage of water within a watershed.

The movement of water in principle is from the higher energy state to lower energy state, and the driving force is the potential different between energy states. The movement of water through the soil is strongly affected by three forces that are a gravitational force, an adhesion force of soil matrix for water and an attraction force of ions and other solutes towards the water. The gravitational force causes water to flow vertically downward due to the gravitational potential energy level of water at a given elevation in the soil profile is higher than that of water at a lower elevation. An adhesion force is referring to an attraction of soil matrix for water, and the force is basically will lead to adsorption and capillarity processes. The capillarity potential refers to the energy state of water molecules adsorbed onto the soil solids. The adhesion forces together with the cohesion forces produce a suction force within the soil that reduces the rate of water movement below the soil surface. Lastly, the attraction of ions and other solutes in water will tend to reduce the energy level in the soil solution.

The infiltration rate decreases as the soil becomes saturated. If the precipitation rate exceeds the infiltration rate, a runoff will usually occur unless there is some physical barrier. The rate of infiltration can be measured using an infiltrometer. There are a few factors control the infiltration rate including soil properties, antecedent water content, and the types of vegetative or ground cover, slope, rainfall intensity and finally movement and entrapment of soil air.

The soil properties are strongly affected by three forces, such as hydraulic conductivity, diffusivity, and water holding capacity. The hydraulic conductivity is of critical importance to infiltration rate since its expresses how easily water flows through soil and is a measure of the soil's resistance to flow. A saturated hydraulic conductivity is referring to the hydraulic conductivity at full saturation. Many of the infiltration equations using the saturated hydraulic conductivity as a parameter since it is easier to determine compared to the unsaturated hydraulic conductivity or the diffusivity. Diffusivity is the proportion of the hydraulic conductivity to the differential water capacity or the flux of water per unit gradient of water content in the absence of other force fields. Since the diffusivity is directly proportional to hydraulic conductivity, usually only the saturated hydraulic conductivity is applied in the approximate infiltration equations. Water holding capacity is the amount of water can hold due to the pore size distribution, texture, structure, the percentage of organic matter, chemical composition and current water content. For saturated condition, the water holding capacity is zero, and the hydraulic head is positive. The soil texture that is consists of the proportion of sand, silt and clay give directly affects the hydraulic conductivity, diffusivity and water holding capacity. Soils with higher sand percentages for example have larger size particle, larger pores, lower water holding capacity and higher hydraulic conductivity, diffusivity and infiltration rates than clay soils which have smaller micropores and bind water molecules more tightly.

Antecedent or initial water content affects the moisture gradient of the soil at the wetting front, the available pore space to store water and the hydraulic conductivity of the soil. Initial water content is, therefore, a critical factor in determining the rate of infiltration and the rate at which the wetting front proceeds through the soil profile. The drier the soil initially, the steeper the hydraulic gradient and the greater the available storage capacity and increase the infiltration rate. The wetting front proceeds more slowly in drier soils because the greater storage capacity, which fills as the wetting front proceeds. Vegetation and ground covers will reduce soil temperature and evaporation process from the soil surface, but at the same time vegetation also losses moisture content through transpiration. Vegetation improves the infiltration rates by loosening soil through root growth. Vegetation along with the natural mulches and plant residues also will intercept raindrops and this process will reduce the bare soil structure damages.

The slope is one of the dominant factors that affect the infiltration rates. A previous study on grass cover slopes was proved a decrease in water infiltration rate was observed with an increase in the slope thickness [16, 17]. Haggard et al. [16] claimed that the slope might have the greatest effect on surface runoff production and infiltration rate when the soil is close to saturation. The infiltration rate of bare soil was observed greater at the sloping condition than in the flat land.

Rainfall intensity is the instantaneous rainfall rate. For a uniform rainfall, the rainfall intensity can be obtained by dividing the depth of rainfall by the duration of rainfall. The maximum rate of the infiltration for the non-ponded condition is known as an infiltration capacity [18]. The rainfall intensity provides the upper limit for the infiltration rate. If the rainfall rate is less than the saturated hydraulic conductivity of the soil, infiltration may continue indefinitely at the rainfall rate without the occurrence of ponding. When the rainfall intensity exceeds the ability of the soil to absorb water, infiltration proceeds at the infiltration capacity. At the time of ponding, the infiltration capacity can no longer keep pace with the rainfall intensity and depression storage fills up and then overflows as runoff. The depression storage will fill faster if the rainfall has higher intensity and time of runoff will occur sooner after the time of ponding.

2.4   HYDROLOGICAL SIMULATION OF THE LANGAT RIVER BASIN

Langat River Basin occupies the south and south-eastern parts of Selangor and a small portion of Negeri Sembilan and Wilayah Persekutuan. The mainstream, Langat River stretches for 180 kilometers and has a total catchment area of 2271 km². The preliminary simulation only focused on the upper part of the catchment area of 305.3 km² as in Figure 2.1, and

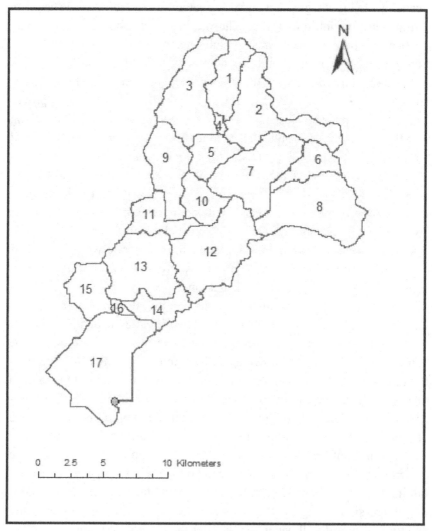

FIGURE 2.1   Watershed delineation output of the study area.

a Kajang streamflow station is selected as the outlet of the water discharge. The study area is covered by five soil types in which the hilly area is covered by a steep land, while rest of it covered by soil types belongs to lowland areas such as Rengam-Jerangau, Munchong Seremban, and Telemong-Akob. The land used maps for the year 2002 was used and reclassify to 15 broad classes of land cover.

The simulation was using 12 years daily rainfall data for the first two years, starting from 1 January 1999 to 31 December 2000 were utilized for the model warm-up, followed by next five years for model calibration and will end by following next five years data validation processes. The SWAT model was utilizing the SCS-Curve Number method in predicting the run-off from daily rainfall. The model is capable to simulate the streamflow of the study area. The calibration output obtained as in Figure 2.2 show simu-lated streamflow is lying above the observed value. A value of R^2 is 0.64 and considered as moderate achievement of the calibration processes. The soil physical input properties as discussed in Section 4.3 are strong needs to be considered in obtaining the required baseflow of the river basin. It can be done by conducting the sensitivity analysis of the input param-eters either manually in the SWAT Model or automated method using the SWAT-CUP program.

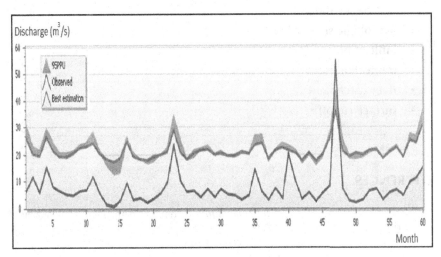

FIGURE 2.2   SWAT five years stream flow simulation output.

2.5 CONCLUSION

The infiltration rates react as one of the major hydrological components that directly influences the runoff capacity of the watershed. The future direction of the study will explore which parameters of physical soil characteristics are the most sensitive to the tropical climate condition, to make recommendation for the best equation to use for each upper catchment soil type and initial water content. The paper will benefit to the soil and conservation engineer by providing the SCS-CN infiltration method in an approximate physical based hydrological model to be used for the river catchment in Malaysia at each antecedent water condition.

ACKNOWLEDGMENT

The project was funded by the Exploratory Research Grant Scheme (ERGS), Ministry of Education, and Universiti Teknologi MARA (UiTM), Malaysia.

KEYWORDS

- hydrologic models
- hydrologic simulation
- infiltration
- infiltration rates
- river catchment
- surface runoff

REFERENCES

1. *Urban Stormwater Management Manual for Malaysia.* Department of Drainage and Irrigation (DID) Malaysia (2000).
2. Liew, Y. S., & Selamat, Z., (2011). Review of Urban Stormwater Drainage System Management. *Malaysia Water Research Journal, 1,* 22–32.

3. Alaghmand, S., Abdullah, R., Abustan, I., Said, M. A. M., & Vosoogh, B. (2012). GIS-Based river basin flood modeling using HEC_HMS and MIKE 11 – Kayu Ara river basin, Malaysia. *JEH*. Vol. 20

4. Mustafa, Y. M., Amin, M. S. M., Lee, T. S., & Shariff, A. R. M. (2011). Evaluation of land development impact on the tropical watershed hydrology using remote sensing and GIS. *Journal of Spatial Hydrology*. *5*(2), 16–30.

5. Supiah, S., Dan'azumi, Salisu, A. R., & Mohamad, A. (2010). Uncertainty analysis of HEC-HMS model parameters for Southern Malaysia. In: The 1st IWA Malaysia Young Water Professionals Conference (IWAYWP, Kuala Lumpur), 2–4 Mac.

6. Julien, P. Y., Ghani, A., Zakaria, N. A., Abdullah, R., & Chang, C. K. (2010). Case Study: Flood Mitigation of the Muda River, Malaysia. *Journal of Hydraulic Engineering @ ASCE*. pp. 251–261.

7. Yusop, Z., Chan, C. H., & Katimon, A. (2007). Runoff characteristics and application of HEC-HMS for modeling stream flow hydrograph in an oil palm catchment. *Water Science & Technology, 56 (8)*, pp. 41–48.

8. Rahim, A. B. E., Yusoff, I., Jafri, A. M., Othman, Z., & Ghani, A. (2012). Application of MIKE SHE modeling system to set up a detailed water balance computation. *Water and Env. Journal*. *26*(4), pp. 490–503.

9. Lai, A. H., & Arniza, F. (2011). Application of SWAT Hydrological Model to Upper Bernam River Basin, Malaysia. *IUP Journal of Env. Sciences*. *5* (2), 7–19.

10. Ayub, K. R., Hin, L. S., & Aziz, H. A. (2009). SWAT Application for hydrologic and water quality modeling for suspended sediments; A case study of St. Langat's Catchment in Selangor. International Conference on Water Resources. pp. 1–7.

11. Beven, K. J. (2000). *Rainfall-Runoff Modeling: The Primer*, John Wiley & Sons, Ltd., New York.

12. Garen, D. C., & Moore, D. S. (2005). Curve Number Hydrology in water quality modeling: uses, abuses, and future directions. *Journal of the American Water Resources Association*. *41*, 377–388.

13. Croley, T. E. II., & He, C. (2005). Great Lakes spatially distributed watershed model of water and materials runoff. *Proceeding, International Conference on Poyang Lake Wetland Ecological Environment. Advanced Workshop on Watershed Modeling and Water Resources Management*. Jiangxi Normal University. Nanchang Jiangxi, P. R. China.

14. Rawls, W. J., Ahuja, L. R., Brakensiek, D. L., & Shirmohammadi, A. (1993). Infiltration and soil water movement. In *Handbook of Hydrology*. McGraw-Hill, Inc.

15. Skaggs, R. W., & Khaleel, R. (1982). *Infiltration In Hydrology of Small Watersheds*. St Joseph, Mich: ASAE.

16. Haggard, B. E., Dawson, B. D., & Brye, K. R. (2005). Effect of slope on runoff from small variable-slope box. *Journal of Environment Hydrology*. *13*, 25.

17. Huat, B. B. K., Ali, H. J., & Low, T. H. (2006). Water infiltration characteristics of unsaturated soil slope and its effect on suction and stability. *Geotechnical and Geological Engineering 24*, 1293–1306.

18. Horton, R. E. (1940). Analysis of runoff plot experiments with varying infiltration capacity. *Transactions of American Geophysicists Union. Part IV*, 693–694.

CHAPTER 3

LANDSLIDE RISK LEVEL ASSESSMENT BASED ON EROSION-INDUCED LANDSLIDE: A CASE STUDY AT UITM CAMPUS, SELANGOR

MOHD FOZI ALI, MUHAMMAD SOLAHUDDEEN MOHD SABRI, KHAIRI KHALID, and NOR FAIZA ABD RAHMAN

Faculty of Civil Engineering, Universiti Teknologi MARA, Shah Alam, Malaysia

CONTENTS

OVERVIEW

Deforestation due to the uncontained development of hill-slope areas is partly the cause of a majority of landslides in Malaysia. Landslides caused losses of lives and thousands to be evacuated when the event takes place and losses in terms of assets. One of the important factors in governing landslide is an erosion-induced landslide. This study focused on the development of landslides risk level based on erosion-induced landslide related to Universal Soil Loss Equation (USLE). The study are involved a collection of secondary data and soil sampling from several of slope and landslide event occurred in Malaysia. Each factor in erosion-induced landslide parameters are provided with an individual scale, and all of the factors are divided into its weight depending on its impact in inducing landslide especially on soil erodibility and rainfall erosivity index. This research assesses a technique of predicting landslide risk level with greater accuracy for future erosion induced landslide. Thus, it will be facilitating landslides detection and can extend its application throughout the country.

3.1 INTRODUCTION

Landslide and slope failure in Malaysia is a serious geologic hazard common to many parts of the world. Landslides caused losses of lives and thousands to be evacuated when the event takes place and not to forget the losses in terms of money. In Malaysia, from 1973 to 2010, over 440 landslides were reported with 31 cases involved fatalities. Also, there are thousands more unreported of minor slope failures and landslides. The landslides at Bukit Antarabangsa in 2008 and Cameron Highland in 2011 involved hillside areas. The second tragedy caused a death of a woman; two people were in critical conditions and ten people missing during the landslide at a Sungai Ruil watershed. The most recent incident of a landslide in Malaysia was incurred at Puchong, Selangor in 2013.

Slope failures due to deforestation for hillside development are partly the main causes for these events. There were some cases where the development projects at hill sites were abandoned for a considerable period, thus affecting the maintenance of the slopes and may cause landslides. The erosion induced landslide poses enormous threats and over the past years as well as the present scenario has caused severe damages [1]. Based on engineering scope, the researcher highlighted that the soil erosion is a process of detachment of soil particles from the soil mass can be a function of rainfall erosivity.

Other researchers were discussed that soil erosion by water agent was one of the problem in tropical countries such as Malaysia, mostly on steep land and in areas devoid of vegetative cover [2]. The researchers also reported that vegetation, soil properties, mechanical properties, rainfall, and slope are interaction factor for soil erosion. Soil erosion can be defined as the transport and detachment of soil particles from the environment and will occur when the intensity of rainfall exceeds the infiltration rate of the disturbed soil [3]. There are two main agents of soil erosion which are wind and water. Also, other researchers emphasized that the process of soil erosion is a long-term and degradation land surface almost invisible which can be caused by wind or water and bad of human activities [4].

The Slope Engineering Branch under the Public Works Department, Malaysia was formed in February 2004 after a rockslide tragedy at KM 21.8 of the New Klang Valley Expressway (NKVE) near Bukit Lanjan on 6 November 2003 which severed the highway and caused massive transportation congestion in Kuala Lumpur for more than six months. Malaysian Government Cabinet Meeting on 26 May 2004 responded to the rising problems and the realization of the impact of landslide hazards in the country, it was decided that study on slope master plan need to be carried out. Considering all these factors, the research was carried out to establish landslide risk level based on erosion-induced landslide that is regarding all the factors governing erosion process.

3.2   METHODOLOGY

The methodology of this study mainly based on laboratory test and data analysis from selected locations and obtained the information that focused on Universal Soil Loss Equation. The data that collected and measured from selected sites and landslide tragedies locations were discussed in terms Universal Soil Loss Equation (USLE). The Universiti Teknologi MARA, Shah Alam, Selangor was selected as a study area of the landslide assessment. The locations of slope were selected based on the potential of a high-risk threat to resident near the sloping areas. Rainfall data were taken from the automatic rainfall station located within the 20 km radius from the study area. The slope parameters that include a slope length and steepness were measured using a Range Finder or as known as TruPulse™

200/200B. Figures 3.1 and 3.2 are showing the location of Universiti Teknologi MARA, Shah Alam, Selangor and a model of Range Finder, respectively.

The particles size distribution, moisture content, and laboratory permeability tests were conducted to determine the soil characteristic that will relate to the (USLE) BS 1377 Parts 1 and 2. The historical landslides events data were collected from few related government or institution agencies namely; Minerals and Geo-science Department Malaysia (JMG); Public

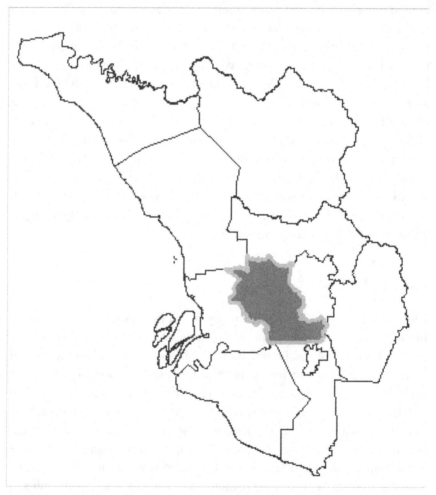

FIGURE 3.1 Location of University Technology MARA, Selangor, Malaysia.

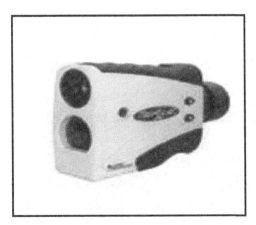

FIGURE 3.2 Range Finder model TruPulse™ 200/200b.

Work Department (Slope Engineering Department, JKR); Department of Irrigation and Drainage Malaysia (JPS) and International Research Centre for Disaster Prevention (IRCDIP). The collected data was based on the Universal Soil Loss Equation (USLE) parameters; rainfall erosion index (R), soil erodibility factor (K), slope length (L), steepness factor (S), vegetative cover factor (C) and erosion control practice factor (P).

Rainfall erosivity index can be measured by using ROSE index that created by [5] and soil erodibility factor can be measured by using ROM scale [1]. Equation (3.1) shows a degree of soil erodibility factor and Figure 3.3 shows the measurement of length and steepness slope.

$$Degree\ Erodibility\ Index = \frac{(\%Sand + \%Silt)}{2(\%Clay)} \qquad (3.1)$$

All these six factors were allocated at a difference weights. The dominants parameters of soil erodibility and rainfall erosivity are weighted as 40% each in the landslide risk nomograph. Both parameters are given a huge impact to erosion-induced landslide, and both are equally important. The other four factors are allocated of 5% weight in the nomograph. Figure 3.4 summarizes these factors in graph formation.

Risk level nomograph is considered both of x-axis and y-axis. The x-axis defines the factor the governing erosion induced landslides. Nomograph

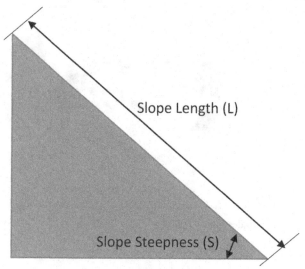

FIGURE 3.3   Measurement of slope length and slope steepness.

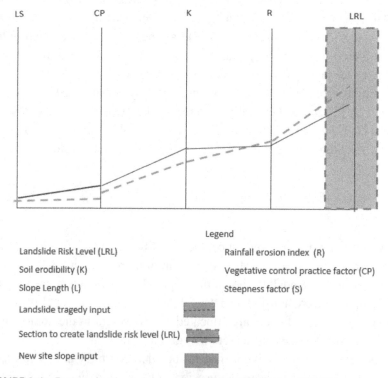

FIGURE 3.4   Proposed estimated nomograph projection based on USLE data.

on y-axis defined the value in percentage factor that consider 5% for which length (*L*) and steepness (*S*), 10% for combination erosion control practice factor (*P*) with vegetative cover factor (*C*). Rainfall erosivity index (*R*) and soil erodibility factor (*K*) is considered 40% weighting for each factor. The value of slope risk level can be created based on the summation of the percentage six (6) factors with a maximum summation of 100%. Finally, the study area can be categories as exposed towards the related landslide using a landslide risk level classification in Table 3.1.

TABLE 3.1   Classification of Landslide Risk Level

Percentage (%)	Landslide Risk Level
< 20.0	Low
20.1–39.9	Moderate
40.0–59.9	High
60.0–79.9	Very high
80.0–100	Critical

3.3   LANDSLIDE RISK NOMOGRAPH FORMULA

Laboratory experiments were conducted to obtain the particle size distribution and hydrometer results. The values of K factor that refers to the ROM scale and the maximum value of the scale is 12. The value of K was obtained and converted to a percentage landslide risk level. Equation (3.2) is used to determine the value of K_n and a maximum percentage of the value of K_n is 40%.

$$K_n = \frac{K}{12} x\, 40\% \qquad (3.2)$$

where, K_n = nomograph percentage ($\leq 40\%$); K = soil erodibility (≤ 12).

The value of rainfall erosivity, R was obtained by calculating the rainfall intensity (I) and rainfall kinetic energy (E). Equations (3.3) and (3.4) are to determine the value of rainfall intensity (I) and rainfall kinetic energy, (E), respectively [7].

$$Ra\ \mathrm{inf}\ all\ Intensity, I = \sum \frac{Amount\ of\ Ra\ \mathrm{inf}\ all}{Period\ of\ Ra\ \mathrm{inf}\ all} \qquad (3.3)$$

$$Ra\ \mathrm{inf}\ all\ Kinetic\ Energy,\ E = 210\ 89\ Log_{10}\ I \qquad (3.4)$$

where, I = rainfall intensity (cm/h).

The value of R factor (ROSE index) has a maximum value of 2000. Referring to Eq. (3.5), other value of R were obtained and converted to a percentage. The maximum percentage of the value of R_n is also 40%.

$$R_n = \frac{R}{2000}\ x\ 40\% \qquad (3.5)$$

where R_n = nomograph percentage ($\leq 40\%$); R = rainfall erosivity index (≤ 2000).

The other factors on landslides risk level are the length of slopes (L) and steepness of the slope (S), which were measured by using the range finder. Referring to Eqs. (3.6) and (3.7), the value of LS_n was obtained and being converted to a percentage [7].

$$LS(\theta) = \left(\frac{\lambda}{72.6}\right)^m \left(65.41\sin^2\theta + 4.65\sin\theta + 0.065\right) \qquad (3.6)$$

where, λ = length of slope (feet); m = 0.5; Θ = degree of slope.

$$LS_n = \left[\frac{LS(\theta)}{LS(90^o)}\right] x\ 10\% \qquad (3.7)$$

where, LS_n = LS nomograph percentage ($\leq 10\%$).

Finally, the landslide risk nomograph can be obtained by a percentage summation of all the parameters. Equation (3.8) has shown the Landslide Risk Nomograph formula.

$$Landslide\ risk\ nomograph\ (\%) = \sum (LS_n + CP_n + K_n + R_n) \qquad (3.8)$$

3.4 RESULTS AND DISCUSSION

A location of a study area is selected based on two main criteria. Since the study area is at UiTM Shah Alam, the location of nearby student residential is preferred, and the sites should be easy to access. In total, five study areas were selected for the UiTM main campus. The soil samplings were made in the selected areas, and raw data were analyzed according to the factors related to erosion to generate the nomograph. Physical properties of the soil were conducted through laboratory experiments such as moisture content and permeability test. Table 3.2 shows the result of soil moisture content and soil permeability for each slope. Based on moisture content results, Jalan Pusat Islam 1 (UiTM), Shah Alam shows the highest percentage of moisture content with 24% and the lowest percentage was 15.4% at Jalan Bangunan Penyelenggaraan 1 (UiTM), Shah Alam.

Table 3.3 shows details USLE parameter outputs of each study area. It was observed a small variation of the slope length and steepness.

TABLE 3.2 Moisture Content and Permeability Test

Location	Area	Moisture Content (%)	Permeability, k (cm/s)	Soil Type
Jalan Pusat Islam 1	1	24	2.00895E-05	Silts
Jalan Pusat Islam 2	2	20	2.00895E-05	Silts
Jalan Bangunan Penyelenggaraan 1	3	15.4	1.39670E-05	Silts
Jalan Bangunan Penyelenggaraan 2	4	21.4	1.35589E-05	Silts
Jalan FSMK 1	5	18	2.75993E-05	Silts

TABLE 3.3 USLE Parameters of the Study Area

Area	Length (m)	Steepness (°)	Cover Practices	K (ROM Scale)	R (ROSE Index)
1	59.5	24.4	0.122	1.19	2878
2	59.5	24.4	0.122	1.24	2878
3	51.8	26.0	0.122	1.39	2878
4	51.8	26.0	0.122	2.24	2878
5	60.0	28.2	0.122	1.25	2878

The ROM scale is highest at Area 4 and minimal in Area 1, and the ROSE index was calculated to be 2878 ton m^2/ha.hr for all the locations.

Table 3.4 has listed the output of each element in landslide risk level for all the study area The maximum risk level for the five (5) slopes at UiTM indicated the "High" landslide risk level. The most possibility of the occurrences were due to rainfall erosivity (R) factor that governing erosion induced landslide where all the locations have a maximum percentage in this factor. The rainfall erosivity factor at UiTM was 2787 ton m^2/ha.hr that has exceeded the maximum for R and maximum value of R was 2000 tons m^2/ha.hr. The landslide risk level nomograph for UiTM Shah Alam was finalized in Figure 3.5.

TABLE 3.4   Landslide Risk Level

Area	LS_n	CP_n	K_n	R_n	LRL	LRL Indicator
1	1.87	1.22	3.97	40.00	4519	High
2	1.87	1.22	4.13	40.00	47.23	High
3	2.09	1.22	4.63	40.00	47.94	High
4	2.09	1.22	7.47	40.00	45.44	High
5	2.40	1.22	4.17	40.00	47.79	High

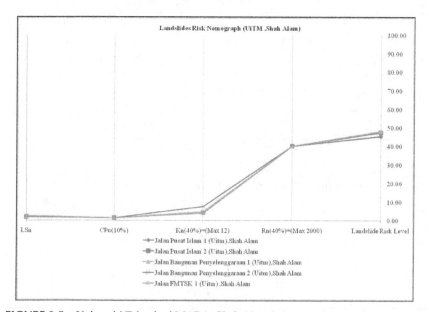

FIGURE 3.5   Universiti Teknologi MARA, Shah Alam Selangor nomograph.

3.5 CONCLUSION

The study was identified the slope risk level based on erosion induced landslide. By applying both nomograph and the developed landslide risk level, the method can be adapted for other potential landslide areas. The method can be used as a tool to develop an erosion risk level guidelines for slopes throughout the nation. The future direction of the study is to include other important factors into the nomograph that will reflect the most precise output of the erosion induced landslide. These parameters including erosion features, type of slope and slope shape.

ACKNOWLEDGMENT

The project was funded by Universiti Teknologi MARA (UiTM), Malaysia. Special thanks to the Malaysian government authorities, Minerals and Geoscience Department Malaysia, Public Work Department and Department of Irrigation and Drainage Malaysia in providing the related data.

KEYWORDS

- **erosion-induced landslide**
- **landslide**
- **landslide risk nomograph**
- **rainfall erosivity**
- **ROSE index**
- **soil erodibility**

REFERENCES

1. Roslan, Z. A. (2009). Professional Lecture: Forecasting Erosion Induced Landslide, *University Publication Centre (UPENA), UiTM.*
2. Ali, M. F., Perumal, T. M., & Salim, S. (2000). The Effect of Slope Steepness on Soil Erosion, Faculty of Engineering, Universiti Putra Malaysia Serdang, Selangor.

3. Eric Goh, & Tew, K. H. (2006). Soil Erosion Engineering, Rill Erosion, University Science Malaysia, pp. 14.

4. Juan, F. S. M., Chris, M. M., Victor, J., & Martin, L. M. (2012). Rainfall kinetic energy–intensity and rainfall momentum–intensity relationships for Cape Verde. Faculty of Geo-Information Science and Earth Observation (ITC), University of Twente, Enschede, The Netherlands.

5. Roslan, Z. A., & Harun, M. S. (2009). *ROSE'* Index For Forecasting Landslide Risk. International Research Centre for Disaster Prevention (IRCDIP), Universiti Teknologi MARA.

6. Software Technical Report: Slope Management and Risk Tracking System, Slope Engineering Branch, Public Works Department, 2005.

7. Wischmeier, W. H., & Smith, D. D. (1978). Predicting Rainfall Erosion Losses – A Guide to Conservation Planning, USDA. *Agriculture Handbook No. 537.*

CHAPTER 4

EFFECT OF CASTING RATE ON THE PERFORMANCE OF NITRATE REMOVAL USING ULTRA-FILTRATION MEMBRANES FOR WASTEWATER REUSE

RAMLAH MOHD. TAJUDDIN,[1]
HASRUL HAFIZUDDIN MUHAMMAD,[2]
TENGKU NURUL AISHAH,[3] and
TENGKU INTAN SURAYA[4]

[1]Faculty of Civil Engineering, Universiti Teknologi MARA, Shah Alam, Selangor, Malaysia

[2]Faculty of Mechanical Engineering, Universiti Teknologi MARA, Shah Alam Malaysia

[3]Malaysian Institute of Transport, Universiti Teknologi MARA, Shah Alam, Selangor, Malaysia

[4]Department of Architecture, Faculty of Built Environment, Universiti Teknologi Malaysia, Skudai, Malaysia

CONTENTS

OVERVIEW

In wastewater, nitrogen compounds as nitrate present at high levels of concentration. The concentration of nitrogen (total as N) in a typical composition of untreated domestic wastewater is in the range of 20 to 85 mg/L [1]. A new regulation on wastewater effluents standards in Malaysia has led to a growth concern in wastewater treatment effluent performance. It is a challenge to produce a new technology with high capability to reduce nutrients especially nitrate on wastewater effluent discharge.

One of the most promising technologies is the membrane filtration process with slow but large impact of success [2, 3]. Membrane technology combined with biological reactors for wastewater treatment has led to the development of membrane bioreactors (MBRs). Ramlah [3] had successfully developed and fabricated asymmetric hollow fiber ultrafiltration (UF) membrane for palm oil mill effluent (POME) tertiary treatment. This reactor was capable to remove nitrate in the range of 30 to 60% removal.

In 1969, Smith et al. claimed that UF membrane system can be used as a replacement for secondary sedimentation tanks in the activated sludge process [4]. Since then, MBRs have been successfully used worldwide in industrial and municipal wastewater treatment in hundreds of applications [5].

Although UF membrane has proven to treat wastewater effectively, several factors can affect the performance of asymmetric membrane. Ismail et al. [6] found that the performance of asymmetric membranes vary depend on a skin layer structure which are subjected to the phase inversion process. In addition to the phase inversion process, it has been recognized that rheological conditions such as casting rate which happened during membrane fabrication will also affect membrane performance by altering molecular orientation [7, 8].

Asymmetric flat sheet ultrafiltration membrane was fabricated from one dope formulation consisting of polysulfone (PSF) as a polymer,

N,N-Dimethylacetamide (DMAc) as its solvent and Polyvinylpyrrolidone (PVP) as a nonsolvent additive. The weight ratio for PSF/DMAc/PVP was carried out at 12/83/5% which is obtained from past research of Chakrabarty (2008). The effect of rheological factor of dope solutions, that is casting rate has been studied in order to fabricate a high membrane performance for nitrate removal. The membranes were fabricated using dry/wet casting technique and then were cast into a flat sheet form with evaporation time in the range of 4 to 8 seconds. Three different casting rate membranes which are 4.19 m/min (DS1), 7.87 m/min (DS2) and 10.38 m/min (DS3) were produced from controlled pressure of 50 kPa, 70 kPa and 90 kPa. Rejection of nitrate were tested using spectrophotometer DR2800 (Method 8039). The effect of operating pressure on membrane performance was also investigated in the range of 2 to 4 bars pressure. The morphology of the prepared membranes was observed using Scanning Electron Microscope (SEM). The experimental results show that casting rate directly influences membrane morphology as well as membrane performance in nitrate rejection test. It was found that the rejection of nitrate increased with increase of casting rate in the following order; DS3>DS2>DS1 and the flux decreased with increase of casting rate in the following order; DS1>DS2>DS3. An increasing trend of flux volume and rejection was observed with increased in operating pressure. The highest rejection of nitrate was found at DS3 with 70.4% rejection at 4 bars pressure.

4.1 INTRODUCTION

In wastewater, nitrogen compounds as nitrate present at high levels of concentration. The concentration of nitrogen (total as N) in a typical composition of untreated domestic wastewater is in the range of 20 to 85 mg/L [1]. A new regulation on wastewater effluents standards in Malaysia has led to a need and growth concern in wastewater treatment effluent performance. It is a challenge to produce a new technology with high capability to reduce nutrients especially nitrate on wastewater effluent discharge.

One of the most promising technologies is the membrane filtration process with slow but large impact of success [2, 3]. Membrane technology combined with biological reactors for wastewater treatment has led to the development of membrane bioreactors (MBRs). Ramlah [3] had

successfully developed and fabricated asymmetric hollow fiber ultrafiltration (UF) membrane for palm oil mill effluent (POME) tertiary treatment. This reactor was capable to remove nitrate in the range of 30 to
60% removal. UF membrane as a replacement for secondary sedimentation tanks in the activated sludge process was first described by Smith [4].
Since then, MBRs have been successfully used worldwide in industrial
and municipal wastewater treatment in hundreds of applications [5].

Although UF membrane has proven to treat wastewater effectively,
several factors can affect the performance of asymmetric membrane. An
asymmetric membrane generally consists of a thin and selective skin layer,
supported on a much thicker layer which is a porous substructure. The surface layer is important to control the separation process and permeation
rates of the membrane while the substructure functions acts as a mechanical support (Baker, 2004). Ismail et al. [6] found that the performance of
asymmetric membranes vary dependent on a thin and defectiveness skin
layer which are subjected to the phase inversion process. In addition to the
phase inversion process, it has been recognized that rheological conditions
such as casting rate which happened during membrane fabrication will also
affect membrane performance by altering molecular orientation [7, 8].

In conventional wastewater treatment plants, nitrogen removal is mostly
achieved by using biological treatment. A number of nitrifying bacteria
are used in the nitrification process. However, this processes demanding
more energy for oxidizing ammonia to nitrite and subsequently to nitrate,
and occupying larger field as well. Therefore, membrane technologies is
being used in wastewater treatment. The UF membrane performance can
be improved by manipulating the rheological factor such as casting rate.
Thus, this paper focused to investigates the performance of asymmetric
flat sheet UF membrane for nitrate removal at different casting rate.

The selection of membrane polymeric materials mainly based on their
characteristics. Chakrabarty [14] has prepared a polysulfone membrane for
bovine serum albumin (BSA) rejection with polyvinyl pyrrolidone (PVP)
was used as additives, and N-methyl–2-pyrrolidone (NMP) and dimethyl
acetamide (DMAC) were used as solvent with 12/5/83 wt%, respectively.
In terms of BSA rejection, the solvent DMAc was found to be more suitable than NMP with the maximum rejection at 76%. Besides, it also
reported the effect of PVP of different molecular weights on the structure

and permeation properties of the prepared membrane. An increase of molecular weight has increased the membrane pore number and pore area as well. It appears that in this research, the effect of rheological conditions such as casting rate were not reported. In view of this, this study attempt to use the material composition based on Chakrabarty [14] with focused on the effect of casting rate on the membrane performance.

4.2 MATERIALS AND METHODS

This section describes the overall process of asymmetric flat sheet ultrafiltration (UF) membrane fabrication. Initially, the formation of asymmetric flat sheet UF membrane includes the dope material composition, preparation of dope solution, membrane casting process, and the testing procedure of the membrane. Figure 4.1 shows the flowchart of the overall experimental works that conducted in this study.

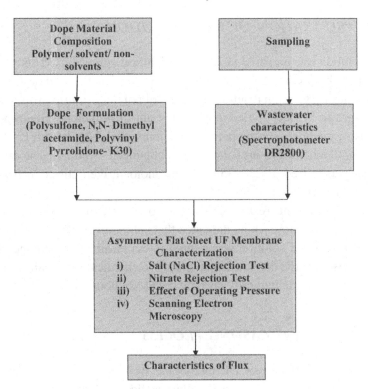

FIGURE 4.1 Flowchart of the overall membrane development and experimental works.

4.2.1   DOPE MATERIAL COMPOSITION

The formulation of dope solution was obtained from past research of Chakrabarty et al. [14]. The dope composition was summarized as in Table 4.1 below.

TABLE 4.1   Dope Materials Composition [14]

Dope Materials	Composition by Weight (%)	Supplier
Polysulfone, PSF (Polymer)	12	ACROS Organic, New Jersey, USA.
N, N-Dimethylacetamide, DMAc (Solvent)	83	ACROS Organic, Geel, Belgium.
Polyvinyl pyrrolidone, PVP K–30 (Additive)	5	ACROS Organic, Geel, Belgium.

4.2.2   PREPARATION OF DOPE SOLUTION

Dope solution was prepared in 1 L for each mixing procedure. This experiment must be prepared in ESCO Ascent Max® Ductless Fume Cabinet because during solution reaction, evaporated polymer solution was smelly and hazard to human. Added a small amount of polymer (PSf) in pellet form using spatula in to the bottom reaction vessel. Mix it with PVP and DMAc until a homogenous solution was obtained (~7 hours). The mixing solution was stirred at 250–350 rpm by motor driven stirrer in heating mantle under temperature of 58°C.

After polysulfone resin were mixed together and polymer solution become crystal yellow which indicated that solution was already homogeneous, the polymer solution was poured into a clean 1 liter storage glass bottle. Subsequently, the bottle then placed in chemical fume cabinet until the polymer solution naturally become cold (~24 hours).

4.2.3   MEMBRANE CASTING PROCESS

The asymmetric flat sheet UF membrane were produced by a dry/wet casting technique using SOLTEQ® Pneumatically Controlled Flat Sheet

Membrane Casting Unit (Model: TR 31-A). At ambient temperature of 27°C, the membranes were cast onto a glass plate with a casting knife notch settling of 150 μm. The membranes were cast at three casting rate by adjusting the casting rate valve pressure while casting speeds were determined using the digital tachometer.

Asymmetric flat sheet membranes were prepared according to the dry/wet phase separation process. The membrane separation process consists of surface treatment and coagulation steps. The surface treatment step was done by allowing the air to be sucked out from the air hood. This step is known as dry phase inversion which involved evaporation process. The surface of the membrane evaporated in the range of 4 to 8 seconds. The evaporation step is considered as essential for the formation of the asymmetric membrane and is critically importance in determining the membrane structures. An increased of evaporation time will affect the membrane, in which considerable loss of solvent occurred from nascent membrane. The coagulation steps involve immersing the membrane sheet into appropriate liquid at appropriate temperature so that the membrane will harden and detach from the casting plate well known as dry/wet inversion phase. In this study, tap water were used as a coagulation medium during this wet phase inversion process.

The glass plate with the flat sheet UF membrane film was immersed into tab water Coagulation Tank. During this process, the solvent in the casting solution (DMAc) was exchanged with a non-solvent (PVP) and phase separation occurs in the film to produce a complete asymmetric membrane structure with a dense top layer and porous sublayer. Finally, the prepared membrane was air-dried for 24 hour at room temperature (27°C).

4.2.4 TEST PROCEDURE

The prepared membranes were tested using test cell (cross-flow method) for in terms of flux and rejection using salt (conductivity meter) and nitrate solution (Method 8039, Spectrophotometer DR2800). Scanning electron microscopic (SEM) then was used to inspect the top surface, porous surface and cross-section of the prepared membrane.

4.3 RESULTS ANALYSIS AND DISCUSSIONS

The membrane used in the experiment was prepared as stated in the methodology. Table 4.2 and Table 4.3 summarizes of the experimental conditions and results obtained from this study.

TABLE 4.2   Summary of the Experimental Conditions

Membrane ID	Casting Rate (m/min)
DS1	4.19
DS2	7.87
DS3	10.38

TABLE 4.3   Summary of the Experimental Results

Membrane	Pressure (bar)	Flux (L/m^2.hr) Salt (NaCl)	(%) Rejection salt (NaCl)	Flux (L/m^2.hr) NO$_3^-$	(%) Rejection NO$_3^-$
DS1	2.0	1.344	7.7	3.09	64
	3.0	2.051	9.1	3.79	65.3
	4.0	2.745	10.67	4.5	67.3
DS2	2.0	1.323	8	2.77	64.6
	3.0	2.002	9.75	3.29	66
	4.0	2.67	11.9	4.38	68.1
DS3	2.0	1.316	10.95	2.61	67
	3.0	1.981	11.59	3.03	67.5
	4.0	2.631	13.36	3.92	70.4

4.3.1 EFFECT OF CASTING RATE ON MEMBRANE PERFORMANCE

At fixed valve pressure of 50, 70 and 90 bar, the casting rate were at 4.19 m/min (DS1), 7.87 m/min (DS2) and 10.38 m/min (DS3), respectively as shown in Table 4.2. These prepared membranes then were used in the nitrate rejection test and SEM.

Before nitrate test was conducted, the salt (NaCl) solution was used to determine the permeation and separation performance of charge solutes and ensure the membrane stability as well. Feed and permeate concentrations of NaCl were measured using a conductivity meter. In every test, the operating pressure were performed at 2–4 bars. According to the experimental results in Table 4.3, the fluxes for NaCl solution test were in the range of 1.344 to 1.316 L/m^2.hr at 2 bars pressure. The fluxes increase with the increased of pressure. At 4 bars pressure, the flux was up to 2.631 L/m^2.hr. For the rejection, the highest NaCl rejection was found at the highest casting rate which is DS3 with 10.95%. At 4 bars pressure, the rejection was increase up to 13.36%. The results appeared to show that the flux were found to decreased with increasing of casting rate while the rejection of NaCl increased with increasing of casting rates.

In the nitrate rejection test, the percentage of nitrate rejection for DS1, DS2 and DS3 were 64%, 64.6% and 67%, respectively with a 2 bars applied pressure. DS3 which is the highest casting rate fabricated membranes in this experiment give the highest removal of nitrate. This result was found similar as NaCl rejection test. The permeate flux for DS1, DS2, and DS3 were about 3.09, 2.77 and 2.61 L/m^2.hr, respectively with a 2 bars applied pressure. An increasing of casting rate has increased the flux permeation. At 4 bars pressure, there was a high increase in nitrate rejection where the percentage was increased about 1.3 to 2.9%. The highest nitrate rejection was at DS3 with 70.4% as shown in Figures 4.2 and 4.3. The results show that the rejection ability was in the following order; DS3>DS2>DS1 which means that the rejection of salt increased with increasing of casting rates.

When the casting rate increased, the pore size decreased to increase selectivity and leading to a better rejection for a solute solution. This is probably due to increased polymer molecules orientation in the membrane active layer. Molecular chains of the prepared membrane with higher casting rate tends to align themselves much better and become closely packed than those experiencing the lower casting rate. This trend was similar with the past findings of Ismail, Chung and Noraini [11–13].

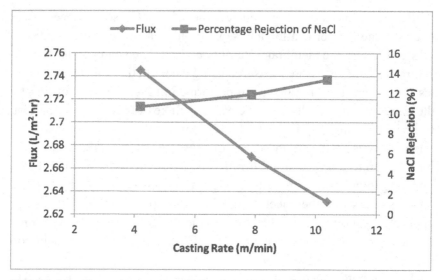

FIGURE 4.2   Flux and percentage of rejection versus casting rate at 4 bars pressure for NaCl solution test.

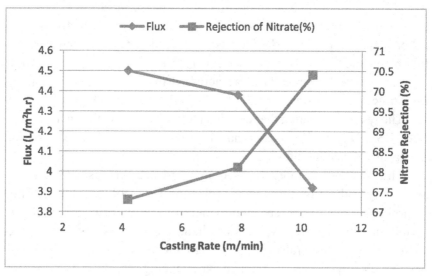

FIGURE 4.3   Flux and percentage of rejection versus casting rate at 4 bars pressure for Nitrate solution test.

4.3.2 EFFECT OF CASTING RATE ON MEMBRANE SURFACE MORPHOLOGY (SEM)

Figure 4.4 represented the SEM for top surface, bottom surface, and cross-section picture of prepared membranes, respectively. It clearly shows that the structural of the top surface membrane is almost similar for all three type of membranes at different casting rate.

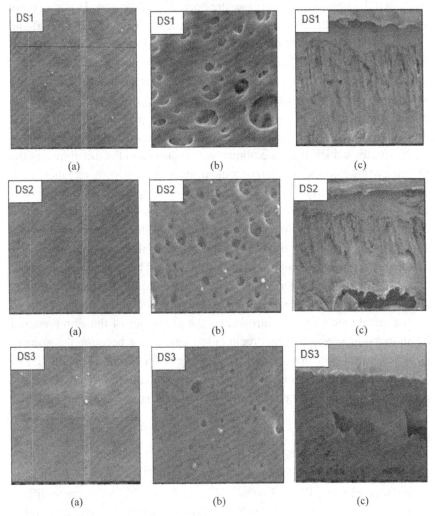

FIGURE 4.4 SEM image at three different casting based on (a) top surface (b) bottom surface and (c) cross-section.

There was no surface pores could be observe on the skin of all three prepared membranes even at magnifications of 30000. The top surface SEM pictures indicate smooth structure devoid of any macroscopic voids for DS2(a). Meanwhile, DS1(a) membrane specimen reveals the presence of circular voids of more or less uniform dimensions on the surface. The surface structure of DS3(a) membrane appears to be intermediate between DS1(a) and DS2(a) with a few voids present in the membrane surface.

When the casting rate increase, the pore size decreased. This statement was proved with the SEM bottom surface picture in DS1(b), DS2(b) and DS3(b). The pore size of prepared membranes were decreased in the following manner: DS1>DS2>DS3. In membrane with higher casting rate, the molecular chains of the solution tend to align themselves much better than those in lower casting rate. This enhanced orientation made the polymer molecules to pack closer to each other. This factor then caused the reduction of membrane pore radius and increased in nitrate rejection.

Normally, a thick layer membrane was observed in the membrane at the lowest casting rate and the skin become thinner as the casting rate increased. The membranes with lower casting rate have more sponge-like substructure. This sponge-like structure was found in DS1(c) and DS2(c). Thus, flux for DS1 was the highest compared to the others. On the contrary, sponge-like structure was not found in DS3(c). This could be the reason for the lower permeate flux in DS3 compared to DS1 and DS2. The sponge-like structure or the microvoids and macrovoids are believed to be formed by the intrusion of non-solvent (PVP) during the wet phase separation. The performance of such a membrane mainly controlled by the properties of the skin prepared by immersion precipitation. Precipitation can occur because the solvent (DMAc) in the polymer solution is exchanged for the non-solvent (PVP) or known as additive. An increased of casting rate caused the polymer molecules to become closely packed and decrease the size of pores and voids.

4.4 CONCLUSIONS

The best performance of fabricated flat sheet UF membrane which prepared from 12 wt.% PSf, 83 wt% DMAc and 5 wt.% PVP at three different casting rate was at DS3 (10.98 m/min). The polymer molecular chains become more aligned and close to each other at this casting rate to give better rejection performance. The highest rejection of nitrate was found at 4 bars pressure with 70.4% removal. This shows that casting rate influence significantly on the morphology and performance of the prepared membranes.

The rejection of nitrate increased with increase of casting rate in the following order; DS3>DS2>DS1 and the flux decreased with increase of casting rate in the following order; DS1>DS2>DS3. When viewed under scanning electron microscopic (SEM), it was proved that the pore size of the prepared membranes was decreased with increase of casting rate.

ACKNOWLEDGEMENT

The authors gratefully acknowledge Faculty of Civil Engineering, Universiti Teknologi MARA, Shah Alam and Institute Molecular Medical Biotechnology (IMMB), Universiti Teknologi MARA Kampus Sungai Buloh for providing the facility to conduct the study. Special thanks to Assoc. Prof. Dr. Ramlah Mohd. Tajuddin for the continuous support and encouragement.

KEYWORDS

- **casting rate**
- **membrane technology**
- **nitrate removal**
- **sedimentation tank**
- **ultrafiltration membrane**
- **wastewater reuse**

REFERENCES

1. SPAN, (2009). *Malaysian Sewerage Industry Guidelines*, Volume IV, Sewage Treatment Plants, 3rd Edition, National Water Services Commission, Malaysia.

2. Baker, R. W., Cussler, E. L., Eykamp, W., Koros, W. J., Riley, R. L., & Strathmann, H. (1991). *Membrane Separation System: Recent Development and Future Directions*, Park Ridge, New Jersey, USA: Noyes Data Corporation, 35–42.

3. Ramlah, M. T. (2006). Development of Novel Ultrafiltration Membrane for Palm Oil Mill Effluent (POME) Membrane Reactor Tertiary Treatment, *Universiti Teknologi Malaysia*, Malaysia.

4. Smith, C. V., Gregorio, D. O., & Talcott, R. M. (1969). The use of ultrafiltration membranes for activated sludge separation, *Proc. 24th Ind. Waste Conf.*, Purdue University, Ann Arbor Science, Ann Arbor, USA, pp. 1300–1310.

5. Bodik, I., Blstakova, A., Dancova, L., & Jakubcova, Z. (2009). Domestic wastewater treatment with membrane filtration – 2 years experience, *Desalination, 240*, 160–169.

6. Ismail, A. F., Hassan, A. R., & Cheer, N. B., (2002). Effect of Shear Rate on The Performance of Nanofiltration membrane for water desalination, *Journal Science Technology, 24*, 879–889.

7. Aptel, P., Abidine, N. A., Ividi, F., & Lafaille, J. P. (1985). Polysulfone Hollow Fibers – Effect of Spinning Conditions on Ultrafiltration Properties. *Journal of Membrane Science 22*, 199–215.

8. Shilton, S. J., Bell, G., & Ferguson, J. (1994). The Rheology of Fiber Spinning And The Properties of Hollow-Fiber Membranes for Gas Separation. *Polymer 35*, 5327–5335.

9. Baker, R. W. (2004). Membrane Technology and Application, *McGraw Hill*, Menlo Park, California, USA, 1st Edition.

10. Chakrabarty, B., Ghoshal, A. K., Purkait, M. K. (2008). Preparation, Characterization, and Performance Studies of Polysulfone Membranes Using PVP as an additive, Department of Chemical Engineering, *Indian Institute of Technology Guwahati*, Guwahati 781039, India.

11. Ismail, A. F., Ng, B. C., & Abdul Rahman W. A. W. (2003). Effects of Shear Rate and Forced Convection Residence Time on Asymmetric Polysulfone Membranes Structure and Gas Separation Performance, *Separation and Purification Technology, 33*, 255–272.

12. Chung, T. S., Wang, K. Y., Matsuura, T., & Guo, W. F., (2004). The Effects of Flow Angle and Shear Rate Within the Spinneret on the Separation Performance of Poly-ethersulfone (PES) Ultrafiltration Hollow Fiber Membranes, *Journal of Membrane Science 240*, 67–79.

13. Nora'aini, A., Sofiah, H., Asmadi, A., & Suriyani, A. R., (2010). Fabrication and Characterization of Asymmetric Ultrafiltration Membrane for BSA Separation: Effect of Shear Rates, *Journal of Applied Science 10* (12), 1083–1089.

14. Chakrabarty, B., Goshal, A. K., & Purkait, M. K., (2008). Preparation, Characterization and Performance Studies of Polysulfone Membranes using PVP As An Additive, Journal of Membrane Science *315*, 36–47.

CHAPTER 5

ADOPTING GREEN AND SUSTAINABLE BUILDING APPROACHES

MUHAMMAD FAIZ MUSA, MOHAMMAD FADHIL MOHAMMAD, MOHD REEZA YUSOF, and ROHANA MAHBUB

Construction Economics and Procurement Research Group, Centre of Studies for Quantity Surveying, Faculty of Architecture, Planning and Surveying, Universiti Teknologi MARA Shah Alam, Malaysia

CONTENTS

OVERVIEW

Green building reduces the negative impact on human health, environment, economy and society during the building lifecycle. Green building overall process starting from the planning, design, construction, operation and maintenance needs to reduce the overall negative impacts on its surrounding.

This study is part of an on-going research on adopting modular construction through the Industrialized Building System (IBS) approach in the Malaysian construction industry. The data and information presented is the review of relevant literatures on this research topic. This study incorporates an analysis of the definitions, benefits and approaches to achieve green building from all over the world. The identification and establishment of a clear definition, benefits and approaches to achieve green building from the analysis are essential so that people and industry players from built and construction environment will be able to understand what is termed as green building.

5.1 INTRODUCTION

Construction industry has significant environmental, social and economic impacts. Building is one of the outputs of the construction industry; largely reflect these impacts during the lifecycle of the building. The positive impacts of the construction industry are: providing buildings and facilities, providing job opportunities (through other industries related to the construction industry) and contributing toward the national economy. Meanwhile, the negative impacts of the construction industry are the noise, dust, and water pollutions including traffic congestion and waste disposal during the construction stage. In addition, a large amount of natural and human resources are also consumed during construction. In addition, completed buildings will continue their impacts towards the environment. According to the World Business Council for Sustainable Development, building block accounts for 40% of total energy consumption [1].

Buildings also produce Green House Gas (GHG) emission which is responsible for global warming. The concentrations of GHG in the atmosphere will be double that of pre-industrial times by as early as 2035. The GHG will subject the Earth to an average temperature rise of over 2 degrees centigrade or, in the worst-case scenario, an increase of 5 degrees centigrade. Such an increase would mean a change in temperatures [2]. The carbon emission of building from worldwide will reach 42.4 billion tones in 2035, adding 43% more of the level in 2007 [3]. In addition, the renovation and refurbishment of building involve the consumption of natural resources and energy, GHG emission, production of noise and pollution. Furthermore, at the end of the life of the building, the disposal of the building is also associated with waste production and energy consumption. The increasing demand of landfill presents a new challenge to all countries that have issues with limited land.

Malaysia is one of the developing countries in South East Asia, has one of the largest carbon footprints and at 37.2 tons of CO_2 equivalent (with land use- change), and ranks 4th in the world in terms of GHG emissions [4]. Such statistics has unsurprisingly led the Malaysian government agenda to create a more sustainable environment by reducing carbon emissions through the application of a green assessment method appropriate to Malaysia climate, environment development context, cultural and social. In the year 2009, Malaysia introduced Green Building Index (GBI) which is Malaysia's green rating tool for towns and buildings. GBI was created to promote sustainability and raise awareness of environmental issue amongst developers, architects, engineers, planners, designers, contractor and the public. For buildings to receive GBI certification, there are six essential criteria, which are (1) Energy Efficiency, (2) Indoor Environmental Quality, (3) Sustainable Site Planning and Management, (4) Materials and Resources, (5) Water Efficiency, and (6) Innovation. In addition, Real Estate and Housing Developers Association Malaysia (REDHA) introduced GreenRE that is the new Malaysian green rating tool. The GreenRE criteria are (1) Energy Efficiency, (2) Water Efficiency, (3) Environmental Protection, (4) Indoor Environment Quality, (5) Other Green Features, and (6) Carbon Emission Development.

5.2 DEFINITION OF GREEN BUILDING

There are many definitions for green building from academician, professionals and organizations from all around the world. GBI defined green building as "Focuses on increasing the efficiency of resource use – energy, water and materials, while reducing building impact on human health and the environment during the building lifecycle, through better design, sitting, construction, operation, maintenance and removal" [5]. Green building has been used as a term interchangeable with high-performance building and sustainable building. According to Robichaud and Anantatmula, there are four pillars of green buildings: (1) enhancing health condition of occupants, (2) the lifecycle consideration during the planning and development process, (3) minimization of impact towards the environment, and (4) the return on investment to the developers [7–9]. Common elements of these definitions are: environmental sustainability, occupants health issues, life cycle perspective and impacts on the community [10]. Table 5.1 highlights the definitions to describe the green building.

There are many different terms and definitions to describe green building. Literatures highlighted several key fundamentals of green building definition. The fundamentals are building lifecycle perspective, human health issues and reducing building impacts toward environment, economic and society.

Based on the reviewed information from relevant literature, a comprehensive definition of green building can be deduced in Figure 5.1:

"Green building reduces the negative impact on human health, environment, economy and society during the building lifecycle. Green building overall process starting from the planning, design, construction, operation and maintenance needs to reduce the overall impacts on its surrounding (human health, environment, economy and society)."

TABLE 5.1 Definitions of Green Building

Authors	Definition of Green Building
Green Building Index (GBI)	Green building focuses on increasing the efficiency of resource use – energy, water and materials while reducing building impact on human health and the environment during the building lifecycle, through better design, sitting, construction, operation, maintenance and removal [5].
US Green Building Council (USGBC)	Green building is significantly to reduce and eliminate the negative impact of buildings on the environment, and the building occupants [6].
Charles J. Kibert	Green building is healthy facilities designed and built in a resource efficient manner, using ecologically based principles [7].
Yanan Li, Li Yang, Baojie He and Doudou Zhao	Green building should be energy saving, land-saving, water saving and material saving, environment benign and pollution reducing [8].

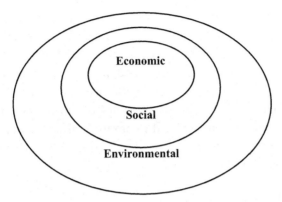

FIGURE 5.1   Three pillars of sustainability.

5.3   SUSTAINABILITY THROUGH GREEN BUILDING

According to the Oxford Dictionaries, sustainability is defined as the capacity to endure. Sustainability requires the reconciliation of environmental, social equity and economic demands which also referred to as the "three pillars" of sustainability. Sustainability in the construction environment is to reduce or eliminate negative impacts from all the activities to the environment for its entire lifetime (from start until the end) while optimizing its economic viability and the human comfort and safety.

5.3.1   ENVIRONMENTAL ASPECTS

Common theme of green building research prioritizes the environmental aspect of sustainability. For example, the recently introduced residential new construction (RNC) launched in July 2013, 80% of the rating tool points are related to environmental sustainability. There are evidences in the existing and current research on environmental sustainability of buildings such as energy efficiency, water efficiency, GHG emission, materials and resources efficiency [11–13]. For example, the structural components of the green building design are added with fly ashes for energy saving and also reduce the waste to the landfill [14]. Similarly, the utilization of precast or prefabrication technologies helps reduce the construction and demolition wastage to a large extent [15]. According to Mohammad, the Industrialized Building System (IBS) construction approach is sustainable, environment friendly, and minimizes wastage and energy efficient [16]. IBS is a construction technique where components are manufactured offsite or in a factory, transported, positioned and assembled with minimal site work [17]. In addition, prefabrication is recognized by both design and construction professionals as one of most common method to prevent injuries particularly related to hazards sustainable elements such as "construction at height, confined places, overhead and with energized electrical system" [18].

5.3.2   ECONOMIC ASPECTS

There are social and economic requirements of green buildings such as access, education, affordability, impacts to the local economy, economic value, cultural perception, inspiration and indoor health [19]. The benefits of energy retrofitting initiatives in a green building reflected through cost savings due to the improved energy efficiency and a potential value added to the property [20]. The energy efficiency contributes to the reduction of the payback period of investment for energy efficiency measures. Even though the initial cost of green building is expensive but due to the cost saving during the building operation, it will benefit the building owners and occupants in a long run.

5.3.3   SOCIAL ASPECTS

Last decades have witnessed the growing concerns on social sustainability in buildings. Construction activities are a social process whereby construction professionals, e.g., Architect, engineer, contractor and others communicate and work together to ensure the completion of a project [21]. In the construction context, social sustainability mainly covers occupational health and safety, the quality of living and future of professional development opportunities [22]. Furthermore, social sustainability in construction means providing a healthy and safe environment to all stakeholders, e.g., Construction personnel, users, and operators should be taken into account during the sustainable design process [23]. In addition, social sustainability should be taken into consideration in the construction projects right from the planning stage [24]. Valdes-Vasquez and Klotz suggested that social sustainability indicators include engaging stakeholders including end users, consideration of the local community and assessment of social impacts.

5.4 BENEFITS OF GREEN BUILDING

There are many researches and studies investigating the costs and benefits associated with green building developments. The aim of the research is to explain the importance of going green which will assist the decision-making process. In addition, there are many benefits associated with green buildings.

5.4.1 ENVIRONMENTAL

From an environmental perspective, green buildings help to improve the urban biodiversity and protect the ecosystem by means of sustainable land use and site management [5, 25]. Reduction of construction and demolition waste is a key component of sustainable building design [26]. Indeed, the recycling rate has to be above 90% in order to mitigate the obvious environmental impacts of demolition and construction waste which means recycled and reused materials in the new buildings [27]. Prefabricated components and modular building systems used in the green buildings will allow the components and systems to be reused, and no waste produced. Compared to conventional buildings, green buildings provide higher performance due to the energy efficiency, carbon emission reduction and water efficiency.

5.4.2 ECONOMIC

Improved building performance, particularly from the lifecycle perspective is associated with the cost savings in green buildings where the operation cost is optimized. Green buildings can save 30% of energy consumption than the conventional buildings [28]. According to Davis Langdon, an additional upfront cost is required for green office building than conventional building [29]. However, the cost of not going green is high as well due to high energy price and carbon trade cost. The cost savings during the operation and maintenance period will help offset the upfront cost required for green building features.

5.4.3   HUMAN ASPECTS

There are studies focus on the human aspects and benefits gained from the green buildings. Human beings are the building occupants and staying in the green buildings for a considerable amount of time.

5.4.4   THERMAL COMFORT

The satisfaction of building users is closely related to the thermal comfort which is complex dynamics of temperature and humidity [30]. This has attracted extensive attention from academician to measure and simulate the thermal comfort level in green building compared to conventional buildings. As a result from the research, the range of room temperature required by building occupants could be proposed [31]. Cultural, behavioral, physiological and psychological factors may play a role as well that attributes to adaptive thermal comfort [32].

5.4.5   INDOOR ENVIRONMENT QUALITY (IEQ)

One of the most critical components of human benefits associated with green building is the indoor environmental quality (IEQ). The IEQ includes the volatile organic compound emissions, and other contaminants are another critical issue in buildings. Therefore, IEQ features in all leading green building rating tools in the world which includes GBI, LEED and Green Mark [33]. Extensive studies have suggested that green building can achieve a higher level of IEQ than conventional buildings, which helps to improve the health and productivity of building occupants. As a result, the level of satisfaction of building user is enhanced. Research conducted by Leaman and Bordass found out that users of green building tend to be more tolerant than those of conventional building in terms of IEQ [34].

5.4.6 HEALTH AND PRODUCTIVITY

Studies also found that the health conditions and level of productivity improved when occupants moved into a green building [35]. Ries et al. [36] suggested that the economic benefits of green building: in terms of productivity and absenteeism should not be overlooked. Their research discovers an increase of 35% of productivity and the absenteeism are reduced when occupants from a conventional building moved into a green building. This health related benefits bring in a broader scope of green building such as social and economic sustainability rather than traditionally environmentally focused [9].

5.5 APPROACHES TO ACHIEVE GREEN BUILDING

There are three categories of critical success factors to achieve green buildings, i.e., managerial, behavioral and technical. These factors are interactive; therefore, a comprehensive consideration is required.

5.5.1 MANAGERIAL

Related organizational and procedural issues are majors barriers for green building developments rather than the lack of innovative technological innovations or rating tools [37]. There are multiple managerial aspects of green buildings, i.e., market level, company level and project level. In market level, it mainly focused on the health of the entire green building market. Similar to other sustainability initiatives, green building is to a large extent dependent on public policies. The purpose of public policy can be either positive or negative incentives (i.e., penalties and compensations). As an initiative of the Council of Australian Government (COAG), the Commercial Building Disclosure (CBD) Program mandates the disclosure of building performance information (e.g., GHG emission and energy efficiency) to building occupants. A Building Energy Efficiency Certificate (BEEC), with information such as the National Australian Built Environment Rating System (NABERS) Energy Star rating, has to be secure prior to the sale, lease and sublease. This certificate is only valid 12 months [38]. The initiative serves as an incentive for Australian building developers and owners to develop more high-performance buildings in the future.

At company level, the implementation of environmental management system (EMS) help to save 90% of energy consumption, reduce 63% of construction and demolition waste, reduces 70% of water consumptions, 80% quality of complaints and lower 20% of accident rate, compared to a company that does not implement EMS [39]. In addition, cost saving is enhanced which then eases the cost management pressure. The commitments from top management are essential for the planning of green building developments [40]. For project level, sets of project management skill are required for managing green buildings. The skills include involvement of specialist consultant of green buildings, adopting green building rating method, providing green building related continuous education and training opportunities to employees and engagement of external stakeholders [9, 41].

5.5.2 CULTURAL AND BEHAVIORAL

The cultural and behavioral factors are also crucial for green building developments [42]. Therefore, it is critical to raise awareness amongst all stakeholders involved in the construction industry on concepts of green building and sustainable development. An experiment to examine the attitude of residents in Hong Kong toward green attributes of residential property [43]. The study discovers the residents prefer to spend more on energy efficiency measures than other green building attributes such as water efficiency, noise control and indoor air quality. Such attitude and behavior of end users is a critical role in promoting green buildings. Therefore, education on green building could be an effective strategy to increase resident awareness of sustainability issues and willingly to pay for green building features.

5.5.3 TECHNOLOGICAL

The utilizations of renewable energy technological innovations are essential to achieve green building objectives and accreditation [44]. The renewable energy is essential due to the depletion of fossil and conventional energy resources and its relation to environmental issues. Therefore, the renewable energy development and utilization of renewable energy in other sectors have priority of many governments. There are a certain number of credits for implementing renewable energy in green building rating tools. The utilization of renewable energy in buildings helps to reduce the energy consumption and emissions. Therefore, the building integrated with renewable energy has become an essential component for green building design and development [45].

Construction and demolition waste management also play a vital role to achieve green building [7]. It is reflected in the green building rating tools. For example, Malaysia GBI RNC rating tool version 3.0, four points will be awarded to the green building for recycled and reused materials. In addition, offsite products and IBS can be reused, recycled and environment friendly [16]. It is a common approach by the government to promote green building materials, innovations and technologies in order to move towards sustainability. The adoption degree of industrialization by Bruno Richard through mechanization and automation can contribute sustainability in construction (e.g., less wastage, high-quality products, mass and efficient production) [46]. Modular Building Institute (MBI) and LEED have acknowledged modular construction, or modular building system is green and sustainable [47].

5.5.4 LIFE CYCLE ASSESSMENT (LCA)

The life cycle assessment (LCA) approach is one of the most popular methods to analyze the technical aspects of green buildings. LCA considers a building as a system while quantifying the material stream and energy consumption flow across various stages of the life cycle. The advantage of LCA is to go beyond the traditional study of focusing on a single stage by extending the investigation to other stages as well (e.g., manufacturing of materials, transportation of materials, energy consumption, GHG emissions during the operation stage and water consumption). Referring to ISO 14040 and ISO 14044, LCA consists of four phases, i.e., impact assessment, goal and scope definition, results interpretation and inventory analysis [48]. The LCA can be applied to the entire building or individual components or materials to evaluate their impacts on the environment thus improve building design [49]. A study on life cycle costing assessment on lighting retrofitting in a one of Malaysia's university as the lighting accounted for 42% of total electricity consumption of buildings. The study discovers that the lighting retrofitting helps to reduce energy consumption by 17–40%. It means a return of investment in 1–2.5 years with consideration of tariff and inflation after using the lighting retrofitting [50].

5.6 CONCLUSION

The current study has highlighted the definition, benefits and approaches to achieve green building. The presented literature has shown the definition, benefits and approaches to achieve green building constitute positive feedbacks and challenges to the implementation of green buildings in Malaysia. Future studies should examine this topic, e.g., Ways to promote green buildings amongst medium and small developers, sustainable assessment tools for infrastructure and transportation networks and others. Finally, the current study is a part of ongoing main research that will further enhance the green building developments through modular construction and modular building system through the IBS approach in Malaysia. The results of the main research will hopefully provide the basis of a guideline to support and enhance the Malaysian construction industry.

KEYWORDS

- **construction industry**
- **green building**
- **industrialized building system (IBS)**
- **modular construction**
- **renewable energy and sustainability**

REFERENCES

1. *Energy Efficiency in Buildings, Business Realities and Opportunities*, WBCSD. (2007).
2. Stern, N., (2006). *The Economics of Climate Change*, Cambridge University Press: UK.
3. Administration, U.S.E.I. (2010). *International Energy Outlook*, U.S. Department of Energy: Office of Integrated Analysis and Forecasting.
4. Pomeroy, J., (2011). *Idea House Future Tropical Living Today*, Singapore: ORO editions.
5. *GBI Explanatory Booklet*, Malaysia: Green Building Index GBI. 12, (2013).
6. *About USGBC, 2004.* cited 2013; Available from: http://www.usgbc.org/about.
7. Kibert, C. J., (2008). *Sustainable Construction: Green Building Design and Delivery.*
8. Li, Y., Yang, L., He, B., & Zhao, D., (2013). *Green Building in China: Needs great promotion.* Sustainable Cities and Society.
9. Robichaud L. B., & Anantatmula, V. S., (2010). *Greening Project Management Practices for Sustainable Construction.* J Manage Eng. *27*(1), 48–57.
10. Zuo, J., & Zhao, Z.-Y., (2014). *Green building research–current status and future agenda: A review.* Renewable and Sustainable Energy Reviews. *30*, 271–281.
11. Pérez, G., Rincoin, L., Vila, A., Gonzalez, J. M., & Cabeza, L. F., (2011). *Green vertical Systems for Buildings as Passive Systems for Energy Savings.* Applied Energy. *88*(12), 4854–4859.
12. Mehta, D. P., & Wiesehan, M., (2013). *Sustainable Energy in Building Systems.* Procedia Computer Science. *19*, p. 628–635.
13. Cheng, C.-L., (2002). *Evaluating Water Conservation Measures for Green Building in Taiwan.* Built and Environment. *38*, 369–379.
14. Drochytka R, Zach, J., Korjenic, A., Hroudova, J., (2012). Improving the Energy Efficiency in Building While Reducing the Waste Using Autoclaved Aerated Concrete Made from the Power Industry Waste. *Energy Build, 58*, p. 319–323.
15. Jaillon, L., Poon, C. S., Chiang, Y. H., (2009). Quantifying the waste reduction potential of using prefabrication in building construction in Hong Kong. *Waste Manage. 29*(1), p. 309–20.
16. Mohammad, M. F., (2013). Construction Environment: Adopting IBS Construction Approach towards Achieving Sustainable Development. in AcE-Bs 2013 Hanoi. Hanoi Architectural University, Hanoi, Vietnam: Elsevier's Procedia Social and Behavioral Sciences.
17. Din, M. I., Bahri, N., Dzulkifly, M. A., Norman, M. R., Kamar, K. A. M., & Hamid, Z. A., (2010). The adoption of Industrialized Building System (IBS) construction in Malaysia: The history, policies, experiences and lesson learned. *International Association for Automation and Robotics in Construction.* p. 8.
18. Dewlaney, K. S., & Hallowell, M., (2012). Prevention Through Design and Construction Safety Management Strategies for High Performance Sustainable Building Construction. *Construction Manage Econ. 30*(2), p. 165–77.
19. Berardi, U., (2013). Clarifying the New Interpretations of the Concept of Sustainable Building. *Sustainable Cities and Society. 8*, p. 72–80.
20. Popescu, D., Bienert, S., Schützenhofer, C., & Bouzu, R., (2012). Impact of energy efficiency measures on the economic value of buildings. *Applied Energy. 89*(1), p. 454–463.

21. Abowitz, D. A., & Toole, T. M., (2009). Mixed method research: fundamental issues of design, validity and reliability in construction research. *Construction Manage Econ. 136*(1), 108–116.

22. Petrovic-Lazarevic, S., (2008). The development of corporate social responsibility in the Australian construction industry. *Construction Manage Econ. 26*(2), 93–101.

23. Wong, K. D., & Fan, Q., (2013). Building information modeling (BIM) for sustainable building design. *Facilities. 31*(3/4), 138–157.

24. Valdes-Vasquez, R., & Klotz, L. E., (2013). Social sustainability considerations during planning and design: A Framework Processes for construction projects. *Construction Manage Econ. 139*(1), 80–90.

25. Henry, A., & Frascaria-Lacoste, N., (2012). Comparing green structures using life cycle assessment: a potential risk for urban biodiversity homogenization. *Int J Life Cycle Assess. 17*(8), 949–950.

26. Yeheyis, M., Hewage, K., Shahria Alam, M., Eskicioglu, C., & Sadiq, R., (2013). An overview of construction and demolition waste management in Canada: a life cycle analysis approach to sustainability. *Clean Technol Environ Policy. 15*(1), 81–91.

27. Coelho, A., & de Brito, J., (2012). Influence of construction and demolition waste management on the environmental impact of buildings. *Waste Manage. 32*(3), 532–541.

28. Economist. (2004). The rise of green building, in *Technology Quarterly.*

29. Langdon, D., (2007). *Cost and Benefit of Achieving Green*, Australia.

30. Zhang, Y., & Altan, H., (2011). A comparison of the occupant comfort in a conventional high-rise office block and a contemporary environmentally concern building. *Built Environment. 46*(2), 535–545.

31. Sicurella, F., Evola, G., & Wurtz, E., (2012). A statistical approach for the evaluation of thermal and visual comfort in free running buildings. *Energy Build. 47,* 402–410.

32. Djongyang, Tchinda, N. R., & Njomo, D., (2010). Thermal comfort: a review paper. *Renew Sustain Energy Rev. 14*(9), 2626–2640.

33. Chuck, W. F., & Kim, J. T., (2011). Building environment assessment schemes for rating IAQ in sustainable buildings. *Indoor Built Environment. 20*(1), 5–15.

34. Leaman, A., & Bordass, B., (2011). Are users more tolerant of green buildings? *Build Res Inform. 35,* 662–73.

35. Gou, Z., Lau, S. S. Y., & Chen, F., (2012). Subjective and objective evaluation of the thermal environment in a three star green office building in China. *Indoor Built Environment. 21*(3), 412–22.

36. Ries, R., Bileca, M. M., Gokhanb, N. M., & Needy, K. L., (2006). The economic benefits of green buildings: a comprehensive case study. *Eng Econ. 51*(3), 259–95.

37. Hakkinen, T., & Belloni, K., (2011). Barriers and drivers for sustainable building. *Build Res Inform. 39*(3), 239–55.

38. CIDB, *The commercial building disclosure: a national energy efficiency program.* 2013; Available from: http://cbd.gov.au.

39. Liu, A. M., Lay, W. S., & Fellows, R., (2012). The contributions of environmental management system towards project outcome case studies in Hong Kong. *Architect Eng Des Manage. 8*(3), 160–169.

40. Beheiry, S. M., W. K. C., Haas, C. T., (2006). Examining the business impact of owner commitment to sustainability. *J Construct Eng Manage. 132*(24), 384–392.

41. GBI, *What is Green Building Index?* (2013). cited 2013; Available from: http://www.greenbuildingindex.org/index.html.
42. Cole, R. J., & Brown, Z., (2009). Reconciling human and automated intelligence in the provision of occupant comfort. *Intel Build Int. 1*(1), 39–55.
43. Chau, C. K., Tse, M. S., & Chung, K. Y., (2010). A choice experiment to estimate the effect of green experience on preferences and willingness to pay for green building attributes. *Build Environ. 45*(11), 25538–25561.
44. Shi, Q., Zuo, J., Huang, R., Huang, J., & Pullen, S., (2013). Identifying the critical factors for green construction – an empirical study in China. *Habitat Int. 40*, 1–8.
45. Hashim, A. H., & Ho, W. S., (2011). Renewable energy policies and initiatives for a sustainable energy future in Malaysia. *Renew Sustain Energy Rev. 15*(9), 4780–4787.
46. Mahbub, R., (2012). Readiness of a Developing Nation in Implementing Automation and Robotics Technologies in Construction: A Case Study of Malaysia. *Journal of Civil Engineering and Architecture. 6*, 858–866.
47. Musa, M. F., Mohammad, M. F., Mahbub, R., & Yusof, M. R., (2014). Enhancing the Quality of Life by Adopting Sustainable Modular Building System in The Malaysian Construction Industry. *Elsevier's Procedia Social and Behavioral Sciences. 153*, 79–89.
48. Dixit, M. K., Fernandez-Solis, J. L., Charles H., & Culp, S. L., (2012). Need for an embodied energy measurement protocol for buildings: a review paper. *Renew Sustain Energy Rev. 16*(6), 3730–3740.
49. Zabalza B., Valero Capilla, I. A., & Aranda Uson, A., (2011). Life cycle assessment of building materials: comparative analysis of energy and environmental impacts and evaluation of the eco-efficiency improvement potential. *Build Environ. 46*(5), 1133–1140.
50. Mahlia, T. M. I., Razak, H. A., & Nursahida, M. A., (2011). Life cycle cost analysis and payback period of lighting retrofit at the University of Malaya. *Renew Sustain Energy Rev. 15*(2), 1125–1132.

CHAPTER 6

FEASIBILITY STUDY ON WIND POWER GENERATION POTENTIAL IN PENINSULAR MALAYSIA

S. S. RAMLI,[1] N. OTHMAN,[1] N. A. KAMARZAMAN,[1] R. AKBAR,[1] Z. FAIZA,[1] and N. MOHAMAD[2]

[1]Faculty of Electrical Engineering, Universiti Teknologi MARA, Pulau Pinang, Malaysia

[2]School of Electrical and Electronic Engineering, Universiti Sains Malaysia, Pulau Pinang, Malaysia

CONTENTS

OVERVIEW

For many decades, fossil fuel such as crude oil, natural gas and coal have been the sources for energy generation. However, these sources will

soon start to diminish while the demand continues to increase, resulting in incessant energy crisis. Therefore, many countries have started to
look at the alternative renewable energy sources such as wind energy to
compensate this decline. Hence, it is essential to determine the wind characteristic for the potential of wind energy generation. This study examines
a feasibility study on wind power generation potential at eleven selected
locations in Peninsular Malaysia based on the real-time wind speed data
obtained from the Meteorological Department (MMD) over a 5-year period
(2005–2009). The analysis was performed using Weibull and Rayleigh
distributions. Characteristics such as the annual and monthly mean wind
speed have been analyzed at three levels of height which were 10 m, 30 m
and 50 m. The results indicate that the highest annual mean wind speed
and wind speed carry the maximum energy estimated at 3.46 m/s in Kota
Bharu followed by Penang and Kuala Terengganu which were 3.13 m/s
and 3.02 m/s at turbine height of 50 m. Generally, the stronger mean wind
speed occurred during the Northeast monsoon (Nov–Mar) and slower mean
wind speed was recorded during the Southwest (Apr–Oct) monsoon at the
islands and east coast areas. Meanwhile, the west part of the Peninsular
Malaysia obtained an almost constant mean wind speed throughout the
year. The annual highest value of the Weibull shape parameter, k and scale
parameter, c were 3.86 and 3.93 m/s, respectively. Meanwhile, the annual
highest value of the Rayleigh parameter, R was 2.8 m/s. At the windiest
location and highest level, the maximum wind power density was found to
be 37.36 W/m^2 while the maximum wind energy density was found to be
327.30 kWh/m^2/year. Due to the lower mean wind speed which was below
than 4.44 m/s at 10 m of height, the wind speed at all selected locations can
be classified as class 1 wind category based on the wind system classification. As a conclusion, most of the selected locations are convenient for the
small-scale wind energy system at the turbine height of 30 m and above.

6.1 INTRODUCTION

Recently, the development of wind energy as one of the renewable energy
sources has grown rapidly in many countries worldwide. At the end of 2001,
the recorded total operational wind power capacity worldwide was around

23,270 MW and distributed to five continents. The highest percentage of installed capacity was about 70.3% in Europe, followed by 19.1% in North America. Meanwhile, Asia and Pacific continents had acquired 9.3% of the total installed capacity. The lower percentages were about 0.9% in the Middle East and Africa and 0.4% in the South and Central America [1].

Various assessments of wind potential have been performed by many countries such as Germany, Holland, Greece, India, China, Turkey, Taiwan and Thailand. Malaysia has the same potential as this country is not only luxuriously endowed with fossil energy resources such as natural gas, coal, and crude oil but it is also rich with renewable energy resources such as solar, wind, biomass, and biogas.

The potential of wind power generation depends on the availability of wind source which varies with location. Thus, it is important to understand the site-specific nature of the wind before wind energy project can be developed [2]. The approach adopted in this study used wind speed data at eleven selected locations in Peninsular Malaysia taken from the Malaysia Meteorological Department (MMD) Station. The recent data of the year 2005 until 2009 obtained are analyzed to identify the location which has the greatest wind power potential in this country.

6.2 MATERIAL AND METHOD

6.2.1 METEOROLOGICAL DEPARTMENT STATION

The wind speed data variations of the year 2005 until the year 2009 were obtained from the Malaysia Meteorological Department (MMD) Station at eleven selected locations namely Langkawi, Alor Star, Penang, Sitiawan, Ipoh, Kuala Lumpur, Kuantan, Kuala Terengganu, Kota Bharu, Mersing and Johor Bharu. The wind speed data were recorded every minute using anemometer while the wind direction data was measured using wind vane. The wind speed was measured in meter per second unit while the wind direction was measured in degree unit. Table 6.1 presents the description of the selected locations in Peninsular Malaysia which comprises of the latitude, the longitudes, the elevation of anemometer (the height of anemometer above ground level) and the density of air at the locations.

TABLE 6.1   Description of Selected Location in Peninsular Malaysia

Locations	Latitude °N	Longitude °E	Elevation (m)	Density of air (kg/m³)
Langkawi	06°33'	99°73'	7	1.224
Alor Star	06°20'	100°42'	5	1.224
Penang	05°23'	100°27'	4	1.225
Sitiawan	04°02'	100°07'	7	1.224
Ipoh	04°57'	101°10'	39	1.220
Kuala Lumpur	03°10'	101°60'	22	1.222
Kuantan	03°78'	103°20'	17	1.223
Kuala Terengganu	05°33'	103°13'	6	1.224
Kota Bharu	06°17'	102°28'	5	1.224
Mersing	02°45'	103°83'	45	1.220
Johor Bharu	01°63'	103°67'	40	1.219

6.2.2   METHODOLOGY

In order to observe the trend and the mean wind speed distribution, the annual and monthly wind speed distribution data at selected locations were analyzed. Since the elevation of anemometer is different for each location, the wind speeds quoted have been adjusted to the same level of height according to the World Meteorological Organization which is about 10 m above the ground level using the Eq. (6.1),

$$\frac{V_1}{V_2} = \left(\frac{H_1}{H_2}\right)^{\alpha} \tag{6.1}$$

where: V_1 = wind speed at height H_1 of 10 m above the ground level; V_2 = wind speed at height H_2 above the ground level; α = power law exponent, which depends on the surface roughness and atmospheric stability.

The method chosen [3] is in accordance with the anemometer height and mean wind speed. The value of α varied from 0.25 to 0.30 [4]. The common heights of the wind turbines, from 30 m to 50 m above the ground level were analyzed in this study [5, 6]. The statistical analysis was conducted using distribution model namely Weibull and Rayleigh distribution. The probability density function, PDF has been determined using Equation 6.2:

$$f(V) = \frac{k}{c}\left(\frac{V}{c}\right)^{k-1} e^{-\left(\frac{V}{c}\right)^{k}} \tag{6.2}$$

where: k = dimensionless Weibull shape parameter; c = scale parameter measured in meter per second unit.

The k values range from 1.5 to 3.0 for most wind conditions [7]. A higher value of k such as 2.5 or 4 indicates that the variation of mean wind speed is small. A lower value of k such as 1.5 or 3 indicates a greater deviation away from mean wind speed [8]. A lower shape factor normally leads to a higher energy production for a given average wind speed [9]. The cumulative distribution function, CDF is given by Eq. (6.3) [5–6, 10–14]:

$$F(V) = 1 - \exp\left(-\left(\frac{V}{c}\right)^{k}\right) \tag{6.3}$$

In this study, the maximum likelihood method was used in the wind speed data analysis. This method was used by [15] in their study for the estimation of parameters of Weibull wind speed distribution for the wind energy utilization purpose. The shape parameter k and the scale parameter c of Weibull distribution with a maximum likelihood method is given by Eq. (6.4) and Eq. (6.5) [6, 16–20]

$$k = \left(\frac{\sum_{i=1}^{n} V_i^k \ln(V_i)}{\sum_{i=1}^{n} V_i^k} - \frac{\sum_{i=1}^{n} \ln(V_i)}{n}\right)^{-1} \tag{6.4}$$

$$c = \left(\frac{1}{n}\sum_{i=1}^{n} V_i^k\right)^{1/k} \tag{6.5}$$

where $V_i (i = 1, 2, 3, n)$ are the observed mean monthly wind speeds and n is the number of nonzero wind speed data points. The Rayleigh distribution is a subset of the Weibull distribution and has one parameter. It is given Eq. (6.6) [21]:

$$f(v) = \frac{V}{c^2} e^{\left(-\frac{V}{2c^2}\right)}$$ (6.6)

The parameter c is given as:

$$c = \frac{v_m}{1.253}$$

where v_m is the mean wind speed.

In the statistical analysis, the EasyFit software has been employed to implement the Weibull distribution. Finally, the wind power density was calculated using Eq. (6.7) [22–26]:

$$\frac{P}{A} = \int_0^\infty P(V) f(V) dV = \frac{1}{2} \rho c^3 \Gamma\left(\frac{k+3}{k}\right)$$ (6.7)

The wind power density indicates how much energy per unit of time is available at the selected area for conversion to electricity by a wind turbine. Since the wind power density is known, the wind energy density for Weibull distribution model can be expressed as Eq. (6.8) [27]:

$$\frac{E}{A} = \frac{1}{2} \rho c^3 \Gamma\left(\frac{k+3}{k}\right) T$$ (6.8)

where T is the time duration which can be 720 hours per month and 8760 hours per year.

6.3 RESULTS AND DISCUSSION

The wind speed distribution analysis has been performed at selected locations in Peninsular Malaysia. The analysis has been classified into two categories which are annual and monthly wind speed distribution. The results of each region have been discussed in detail. The potential location to apply wind energy system has also been identified.

6.3.1 ANNUALLY MEAN WIND SPEED DISTRIBUTION

The wind speed data utilized in this study was the hourly wind records over a five years period (2005 to 2009) from eleven stations. The primary data of annual mean wind speed is shown in Table 6.2. Based on Table 6.2, it is seen that the highest average annual mean wind speed was about 2.70 m/s in Mersing followed by 1.98 m/s in Kota Bharu. On the other hand, Sitiawan acquired the lowest average annual mean wind speed which was approximately 0.83 m/s. In addition, the average annual mean wind speed at other areas was between 1.18 m/s and 1.85 m/s. Since the elevation of the anemometers is different for each location, the wind speeds quoted are all corrected to 10 m, 30 m and 50 m according to the Eq. (6.1)

The results of annual mean wind for the three levels of height are presented in Tables 6.3–6.5. From Table 6.3, it is seen that the annual mean wind speed occurred every year at each location starting at an almost similar speed. Besides, the highest average annual mean wind speed was about 2.36 m/s in Kota Bharu, followed by Penang at 2.12 m/s and Langkawi at 2.03 m/s. This is due to the elevation of speed in Mersing where 45 m has

TABLE 6.2 Primary Data of Annual Mean Wind Speed

Location/Year	Wind speed (m/s)					Mean wind speed (m/s) (2005–2009)
	2005	2006	2007	2008	2009	
Langkawi	2.03	1.83	1.75	1.80	1.85	1.85
Alor Star	1.62	1.51	1.49	1.49	1.30	1.48
Penang	1.77	1.69	1.71	1.63	1.62	1.68
Sitiawan	0.77	0.65	0.95	0.93	0.87	0.83
Ipoh	1.51	1.56	1.54	1.49	1.49	1.52
Kuala Lumpur	1.59	1.55	1.70	1.62	1.34	1.56
Kuantan	1.28	1.43	1.34	1.28	1.28	1.32
Kuala Terengganu	1.62	1.77	1.84	1.73	1.75	1.74
Kota Bharu	2.12	1.97	2.05	1.95	1.80	1.98
Mersing	2.69	2.67	2.60	2.81	2.74	2.70
JohoreBharu	1.10	1.07	1.20	1.22	1.29	1.18

TABLE 6.3   Annual Mean Wind Speed At 10 m Height

Location/Year	Wind speed (m/s)					Mean wind speed (m/s) (2005–2009)
	2005	2006	2007	2008	2009	
Langkawi	2.22	2.01	1.92	1.97	2.02	2.03
Alor Star	1.92	1.80	1.77	1.76	1.72	1.79
Penang	2.23	2.13	2.15	2.05	2.04	2.12
Sitiawan	0.84	0.71	1.04	1.01	0.95	0.91
Ipoh	1.07	1.11	1.09	1.06	1.06	1.08
Kuala Lumpur	1.31	1.28	1.40	1.33	1.09	1.28
Kuantan	1.13	1.25	1.17	1.12	1.12	1.16
Kuala Terengganu	1.84	2.01	2.10	1.97	1.98	1.98
Kota Bharu	2.53	2.35	2.44	2.32	2.14	2.36
Mersing	1.84	1.84	1.78	1.93	1.88	1.85
Johor Bharu	0.78	0.75	0.85	0.86	0.91	0.83

TABLE 6.4   Annual Mean Wind Speed at 30 m Height

Location/Year	Wind speed (m/s)					Mean wind speed (m/s) (2005–2009)
	2005	2006	2007	2008	2009	
Langkawi	2.93	2.64	2.52	2.59	2.66	2.67
Alor Star	2.52	2.37	2.34	2.32	2.26	2.36
Penang	2.93	2.80	2.84	2.69	2.69	2.79
Sitiawan	1.10	0.93	1.37	1.34	1.25	1.20
Ipoh	1.41	1.46	1.44	1.40	1.39	1.42
Kuala Lumpur	1.72	1.68	1.84	1.76	1.44	1.69
Kuantan	1.48	1.65	1.54	1.48	1.47	1.52
Kuala Terengganu	2.42	2.65	2.76	2.59	2.61	2.61
Kota Bharu	3.33	3.09	3.21	3.05	2.82	3.10
Mersing	2.43	2.42	2.35	2.54	2.48	2.44
Johor Bharu	1.02	0.99	1.12	1.14	1.21	1.10

been decreased to the standard height of 10 m while the elevation of speed in Kota Bharu which was 5 m has also been increased to the same level of height. The lowest average annual mean wind speed was around 0.83 m/s

TABLE 6.5 Annual Mean Wind Speed at 50 m Height

Year	2005	2006	2007	2008	2009	Mean wind speed (m/s) (2005–2009)
Location			Wind speed (m/s)			
Langkawi	3.33	3.00	2.86	2.94	3.03	2.96
Alor Star	2.87	2.69	2.65	2.63	2.57	2.64
Penang	3.33	3.18	3.22	3.06	3.05	3.13
Sitiawan	1.25	1.06	1.55	1.52	1.42	1.39
Ipoh	1.60	1.66	1.63	1.59	1.58	1.62
Kuala Lumpur	1.96	1.91	2.09	1.99	1.64	1.91
Kuantan	1.68	1.87	1.75	1.68	1.67	1.74
Kuala Terengganu	2.75	3.01	3.13	2.95	2.97	3.02
Kota Bharu	3.78	3.51	3.65	3.46	3.20	3.46
Mersing	2.76	2.75	2.67	2.89	2.81	2.77
JohoreBharu	1.16	1.13	1.27	1.29	1.37	1.27

in Johor Bharu and the other locations have had an average annual mean wind speed between 0.91 m/s to 1.79 m/s. The result of mean wind speed at 10 m is found to be in opposition to the primary data.

The result of annual mean wind speed at 30 m height is presented in Table 6.4. It is evident that the wind speed has become stronger as elevation of anemometer increased from 10 m to 30 m height. Thus, the strongest average annual mean wind speed was the increase from 2.36 m/s at 10 m height to 3.10 m/s at 30 m height. Meanwhile, the lowest average annual mean wind speed was about 1.10 m/s.

Later, the elevation of anemometers has been increased to 50 m height. Consequently, the average annual mean wind speed has continuously increased. Thus, the windiest location which is Kota Bharu has had an average annual mean wind speed of around 3.46 m/s followed by 3.13 m/s in Penang and 2.96 m/s in Langkawi. The lowest average annual mean wind speed also increased from 0.83 m/s at 10 m height to 1.27 m/s at 50 m height. From the overall results, it can be concluded that the annual mean wind speed increases as the elevation of anemometers increases from 10 m to 50 m height.

6.3.2 MONTHLY MEAN WIND SPEED

Besides performing the annually mean wind speed analysis, the monthly mean wind speed analysis has also been executed to observe the pattern of mean wind speed and determine which month has stronger and weaker wind speed. Similar to the annual mean wind speed, the primary data of monthly mean wind speed has also been adjusted to 10 m, 30 m and 50 m height. The results are illustrated in Figures 6.1–6.4.

The result shows that the monthly mean wind speed was affected by geographical and topographical descriptions of the location. The stronger mean wind speed in Langkawi occurred during the dry season when the winds from the Northeast dominate the region starting from December to March. The range of monthly mean wind speed at this location was between 2.21 m/s in March to 2.96 m/s in January. Meanwhile, the weaker wind speed occurred during the Southwest monsoon from April to October. The range of monthly mean wind speed during the monsoon was between 1.62 m/s in May; and 1.88 m/s in August. The factor contributing to this phenomenon is the decreasing temperature during dry and wet seasons. Even though such a decrease was expected, it has caused thermal convection which resulted in some of the momentum of the upper air to be transmitted to the surface layers. This has caused the noticeable increase in the previously mentioned monthly mean wind speed. In addition, the monthly mean wind speed in Langkawi increased simultaneously as the elevation increases from 10 m to 30 m and 50 m height. At 30 m height, the range of stronger mean wind speed during the dry season varied from 2.90 m/s to 3.90 m/s while during the wet season, it was between 2.14 m/s and 2.47 m/s. At 50 m height, the range of monthly mean wind speed increased to between 3.30 m/s to 4.43 m/s during windy season and 2.43 m/s to 2.81 m/s during calmer season.

The monthly mean wind speed in Penang illustrates that the trend of mean wind speed was almost similar with Langkawi Island at which the mean wind speed started to become stronger from December until March. At 10 m height, the monthly mean wind speed varied from 2.13 m/s to 2.96 m/s. During this time, the Northeast monsoon dominated the island. The weather was warm. Meanwhile the mean wind speeds slowly decreases and became stable from April to November with the range of 1.79 m/s to 1.99 m/s. During these months, the island has been dominated

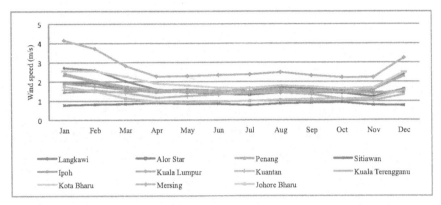

FIGURE 6.1   Primary data of monthly mean wind speed.

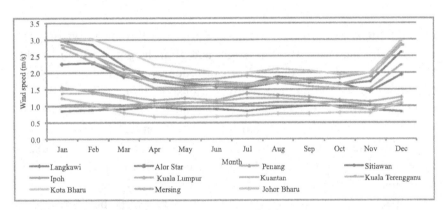

FIGURE 6.2   Monthly mean wind speeds at 10 m height.

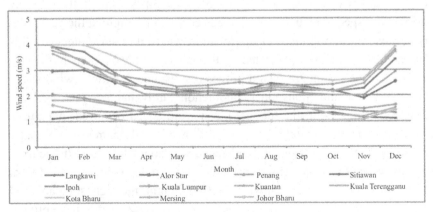

FIGURE 6.3   Monthly mean wind speeds at 30 m height.

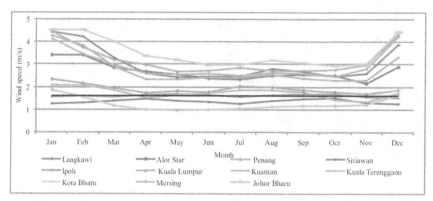

FIGURE 6.4   Monthly mean wind speeds at 50 m height.

by the Southwest monsoon. Meanwhile, the monthly mean wind speed dramatically increased at 30 m height and 50 m height. At 30 m height, the range of wind speed during Northeast monsoon varied from 2.80 m/s to 3.90 m/s while during the Southwest monsoon, it varied from 2.35 m/s to 2.62 m/s. At 50 m height, the range continued to increase to 3.18 m/s to 4.43 m/s on dry season and 2.68 m/s to 2.98 m/s on wet season.

Besides that, the wind speed at Sitiawan was sluggish and did not experience the monsoon seasons. This is because, the location of Sitiawan is not in the wind path and it is also hidden by the hills. Most of the mean wind speed at 10 m height was below than 1.0 m/s while but increased to the range of 1.09 m/s to 1.30 m/s at 30 m height and continued to increase to the range of 1.24 m/s to 1.47 m/s at 50 m height. Moreover, the monthly mean wind speed in Kuala Lumpur was constant throughout the year. This is because the topography of the location where it is surrounded by forests and hills. It is also protected by the Titiwangsa Mountains in the east and Indonesia's Sumatra Island in the west. Hence, the mean wind speed became lower and monsoon phenomenon rarely happens at this location. At 10 m height, the highest mean wind speed occurred in January which was approximately 1.56 m/s, followed by 1.45 m/s in February. Conversely, the lowest mean wind speed was about 1.13 m/s. It occurred in November. Other months were experiencing mean wind speed between 1.17 m/s and 1.37 m/s. Meanwhile at 30 m height, the highest mean wind speed was about 2.34 m/s in January followed by 2.16 m/s in February. On the other hand, the lowest mean wind speed at this level of height was about 1.49 m/s in November. At 50 m height, the monthly mean wind speed has risen every month. As the result, the highest mean wind speed was about 2.34 m/s in January and the lowest was around 1.70 m/s in November. Whilst, other months have obtained mean wind speed in the range of 1.75 m/s to 2.05 m/s.

Besides that, the east coastal area which is Kota Bharu obtained the strongest monthly mean wind speed all over the year than other locations. Basically, the strongest wind speed appeared in Kota Bharu for the period of Northeast monsoon starting from November until March. During this time, the wind from South China Sea blew and dominated the region since there was no impediment such as hills or mountains. It was expected that this region received heavy rainfall. The season in Kota Bharu is in contradictory to the Langkawi and Penang Islands at which Kota Bharu was experiencing the wet season when the islands were experiencing the dry season. At 10 m height, the mean wind speed during Northeast monsoon was between 2.69 m/s and 3.03 m/s. Moreover, the months of January and February obtained the same value of mean wind speed which was approximately 3.03 m/s. On the other hand, the mean wind speed in December was about 2.97 m/s. Moreover the mean wind speed was frail from April to November with the range of 1.95 m/s to 2.26 m/s. The wind speed became worse in these months. At 30 m height, January and February months witnessed the highest mean wind speed as much as 3.99 m/s while August as the calmest month had the lowest mean wind speed at around 2.57 m/s. Meanwhile, at 50 m height, the strongest mean wind speed continuously soars from 3.99 m/s to 4.53 m/s while the lowest mean wind speed also rose from 2.57 m/s to 2.92 m/s.

Furthermore, Kuala Terengganu has also enjoyed the same season as Kota Bharu. However the duration of wet season in Kuala Terengganu is shorter than Kota Bharu. The wet season occurs from December to February while the dry season occurs from March to November. During the wet season, the mean wind speed at 10 m height was above 2 m/s. The mean wind speed in December was approximately 2.82 m/s followed by January, 2.76 m/s and February at 2.32 m/s. There is a regular mean wind speed throughout the wet season. However, during the dry season, the monthly mean wind speed was below 2 m/s. At 30 m height, the monthly mean wind speed soared especially during the wet season which was about 3.71 m/s in December, 3.63 m/s in January and 3.06 m/s in February. On the other hand, the monthly mean wind speed was in the range of 2.16 m/s and 2.62 m/s throughout the dry season. At 50 m height, the monthly mean wind speed increased to 4.22 m/s, 4.12 m/s and 3.47 m/s during the windiest months. Meanwhile, the monthly mean wind speeds were still above 2 m/s during the calmest months at which the range was from 2.46 m/s to 2.98 m/s.

The final selected location, Johor Bahru obtained the lowest monthly mean wind speed. At 10 m height, the highest mean wind speed occurred in January which was about 1.24 m/s. Then it is followed by December (1.09 m/s) and February (1.05 m/s). However other months obtained the monthly mean wind speed below than 1.0 m/s. At 30 m height, the wind speed increased from 1.24 m/s to 1.63 m/s in January. While December and February have had the monthly mean wind speed at around 1.43 m/s and 1.39 m/s, respectively. Meanwhile, the monthly mean wind speed in the other months was close to 1.0 m/s. At 50 m height, the wind speed increased to 1.85 m/s in January, 1.63 m/s in December and 1.58 m/s in February. On the other hand, other months had obtained the mean wind speed above 1.0 m/s.

From the results, it can be concluded that the stronger mean wind speed occurred during the Northeast monsoon and lower one during the Southwest monsoon at the islands and east coast areas. Meanwhile, the mean wind speed was almost constant throughout the year at other locations in the west part of Peninsular Malaysia. Besides, the monthly mean wind speed increased as the elevation increases from 10 m height to 50 m height. At 30 m and 50 m height, the monthly mean wind speed at all selected locations were above 2 m/s except Sitiawan and Johor Bharu. Generally, the cut in speed of wind turbine was 2 m/s. As a result, it is recommended that the minimum elevation of anemometer should be 30 m height above the ground level so that the small scales wind turbine such as 1 kW or 2 kW vertical axis wind turbine can be used to capture the stronger wind speed and generate the higher energy.

6.3.3 STATISTICAL ANALYSIS

In Weibull distribution model, there are two main parameters that have been calculated which are shape parameter, k and scale parameter, c. The results of Weibull distribution, Rayleigh distribution, and maximum of wind speed at the three levels of height are shown in Table 6.6.

TABLE 6.6 Weibull, Rayleigh Distribution and Maximum of Wind Speed at 10 m, 30 m and 50 m Height

Height	10 m				30 m				50 m			
Location	k	c	V_{MAXE}	R	k	c	V_{MAXE}	R	k	c	V_{MAXE}	R
Langkawi	2.92	2.27	2.71	1.62	2.92	3.00	3.59	2.14	2.92	3.40	4.07	2.42
Alor Star	3.86	1.98	2.21	1.43	3.86	2.61	2.91	1.88	3.86	2.97	3.31	2.14
Penang	3.19	2.37	2.76	1.69	3.19	3.12	3.63	2.23	3.19	3.54	4.12	2.53
Sitiawan	3.23	1.03	1.20	0.74	3.21	1.35	1.57	0.97	3.03	1.54	1.82	1.10
Ipoh	3.86	1.19	1.33	0.86	3.86	1.56	1.74	1.13	3.86	1.77	1.97	1.28
Kuala Lumpur	3.43	1.43	1.63	1.03	3.43	1.88	2.15	1.35	3.43	2.14	2.45	1.54
Kuantan	3.34	1.29	1.48	0.92	3.34	1.70	1.96	1.22	3.34	1.93	2.22	1.38
Kuala Terengganu	2.27	2.23	2.95	1.58	2.27	2.94	3.88	2.08	2.27	3.34	4.41	2.36
Kota Bharu	2.96	2.63	3.13	1.87	2.96	3.46	4.12	2.46	2.96	3.93	4.68	2.80
Mersing	3.45	2.06	2.35	1.48	3.45	2.72	3.11	1.95	3.44	3.08	3.52	2.21
Johor Bharu	2.40	0.94	1.21	0.67	2.40	1.18	1.52	0.83	2.40	1.35	1.74	0.96

Referring to the Weibull parameters results in Table 6.6, it is found that the values of shape parameter, k are generally higher than those of scale parameter, c for all locations. Moreover, the values of k are identical at which the value of k varies from 2.27 to 3.86 for all locations. However, the values of c are slightly different. It depends on the height of mast measurement. The parameter c increased as the elevation increases. Besides that, the maximum mean wind speed, V_{MAXE} has also been calculated once the value of k and c is given. Likewise the c parameter results, the value of V_{MAXE} increased as the height of mast measurement increases. The maximum mean wind speed, V_{MAXE} carry the maximum amount of energy. The results of Weibull distribution are compared with Rayleigh distribution model. This is important to determine which distribution described the actual data better. From this table, it is apparent that the values of Rayleigh distribution increased as the elevation increases. The highest values of Rayleigh were obtained at windiest location which is Kota Bharu. Conversely, the minimum values were obtained in Johor Bharu as the lowest wind speed regime.

6.3.4   WIND POWER DENSITY AND WIND ENERGY DENSITY

In order to determine how much energy per unit of time is available at the selected area for conversion to electricity by a wind turbine, the wind power density has been determined. The results of annual mean wind power density at 10 m, 30 m and 50 m height are illustrated in Table 6.7.

From Table 6.7, it is understood that the highest annual mean wind power density at 30 m height is approximately 25.50 W/m² which is located in Kota Bharu, while the lowest of annual mean wind power density is about 1.14 W/m² in Johor Bharu. The results are mainly affected by the wind speed characteristic due to the fact that the wind power density is proportional to the cube of the wind speed. Obviously, the higher and better results can be obtained at higher wind speed conditions. Besides, the annual mean wind power density has sharply increased as the elevation increases. It can be concluded that the higher wind power density can be achieved at windier location and it increases as the elevation increases.

TABLE 6.7   Annual Mean Wind Power Density at 10 m and 30 m HEIGHTS

Location/Height	WPD (W/m²)			WED (kWh/m²/year)		
	10 m	30 m	50 m	10 m	30 m	50 m
Langkawi	7.24	16.72	24.34	63.46	146.47	213.22
Alor Star	4.40	10.07	14.84	38.52	88.22	129.99
Penang	7.95	18.15	26.51	69.67	158.96	232.18
Sitiawan	0.65	1.47	2.23	5.69	12.85	19.50
Ipoh	0.95	2.15	3.14	8.36	18.84	27.51
Kuala Lumpur	1.71	3.88	5.72	14.95	33.96	50.09
Kuantan	1.26	2.89	4.23	11.06	25.32	37.04
Kuala Terengganu	8.02	18.38	26.95	70.28	161.04	236.12
Kota Bharu	11.20	25.50	37.36	98.09	223.35	327.30
Mersing	5.09	11.72	17.03	44.60	102.67	149.20
Johor Bharu	0.58	1.14	1.71	5.05	9.98	14.94

Another outcome of the statistical analysis is the wind energy density. The wind energy density and the wind power density have been computed and the result is as shown in Table 6.7. Based on this table, it is seen that the wind energy density is proportional to the wind power density. The maximum value of annual wind energy density in Kota Bharu is approximately 327.30 kWh/m²/year. Nevertheless, the minimum value of annual wind energy density is obtained from Johor Bharu which is approximately 14.94 kWh/m²/year. Besides, the results of annual wind energy density are also comparative to the height of the mast measurement. It indicates that when the elevation increases, the annual wind energy density increases too.

6.4   CONCLUSION

The feasibility study on wind power generation potential in Peninsular Malaysia areas has been a success. Wind speed distribution analysis showed that the mean wind speed increases as the height of mast measurement increases. Besides, the wind speed was stronger during the Northeast monsoon at all selected locations. Furthermore, the results of shape factor, k is identical at 10 m, 30 m and 50 m height and the scale parameter, c is slightly different depending on the height of mast measurement. It is also higher at the windier location and lower at the calmer regions. The result reveals that the highest annual mean wind power density is about 37.36 W/m² in Kota Bharu, while the lowest is about 1.71 W/m² in Johor Bharu. According to the wind system classification, the wind speed characteristic at all selected locations in Peninsular Malaysia is classified as class 1 wind category at which the annual mean wind speed is ≤4.44 m/s and the wind power density is ≤100 W/m² at 10 m height.

KEYWORDS

- alternative renewal energy
- energy generation
- fossil fuel
- wind energy
- wind power generation
- wind speed

REFERENCES

1. Ackermann T., & Söder, L. (2000). Wind energy technology and current status: a review, *Renewable and Sustainable Energy Reviews, 4(4)*, 315–374.
2. Sopian K., Othman M., Yatim B., & Daud, W. (2005). Future directions in Malaysian environment friendly renewable energy technologies research and development, *Science and Technology Vision, 1,* 30–36.
3. Justus, C. G., & Mikhail, A. (1976). Height variation of wind speed and wind distributions statistics, *Geophysical Research Letters, 3,* 261–264.
4. Exell, R. H. B., & Fook, C. T. (1986). The wind energy potential of Malaysia, *Solar Energy, 36,* 281–289.
5. Li, M., & Li, X. (2005). Investigation of wind characteristics and assessment of wind energy potential for Waterloo region, Canada, *Energy Conversion and Management, 46,* 3014–3033.
6. Ahmed Shata, A., & Hanitsch, R. (2006). Evaluation of wind energy potential and electricity generation on the coast of Mediterranean Sea in Egypt, *Renewable Energy, 31,* 1183–1202.
7. Kavak Akpinar, E., & Akpinar, S. (2005). A statistical analysis of wind speed data used in installation of wind energy conversion systems, *Energy Conversion and Management, 46,* 515–532.

8. Bhattacharya P., & Bhattacharjee, R. (2009). A study on Weibull distribution for estimating the parameters, *Wind Engineering, 33*, 469–476.

9. van Alphen, K. van Sark, W. G., & Hekkert, M. P. (2007). Renewable energy technologies in the Maldives—determining the potential, *Renewable and Sustainable Energy Reviews, 11*, 1650–1674.

10. Celik, A. N. (2007). A techno-economic analysis of wind energy in southern turkey, *International Journal of Green Energy, 4*, 233–247.

11. Weisser, D. (2003). A wind energy analysis of Grenada: an estimation using the 'Weibull' density function, *Renewable Energy, 28*, 1803–1812.

12. Darwish, A., & Sayigh, A. (1988). Wind energy potential in Iraq, *Solar & Wind Technology, 5*, 215–222.

13. Ozerdem, B. Ozer, S., & Tosun, M. (2006). Feasibility study of wind farms: A case study for Izmir, Turkey, *Journal of Wind Engineering and Industrial Aerodynamics, 94*, 725–743.

14. Karsli V., & Geçit, C. (2003). An investigation on wind power potential of Nurdağı-Gaziantep, Turkey, *Renewable Energy, 28*, 823–830.

15. Stevens M., & Smulders, P. (1979). The estimation of the parameters of the Weibull wind speed distribution for wind energy utilization purposes, *Wind Engineering, 3*, 132–145.

16. LunI. Y., & Lam, J. C. (2000). A study of Weibull parameters using long-term wind observations. *Renewable Energy, 20*, 145–153.

17. Ramírez P., & Carta, J. A. (2006). The use of wind probability distributions derived from the maximum entropy principle in the analysis of wind energy. A case study. *Energy Conversion and Management, 47*, 2564–2577.

18. Merzouk, N. K. (2000). Wind energy potential of Algeria, *Renewable Energy, 21*, 553–562.

19. AkdağS. A., & Güler, Ö. (2010). Evaluation of wind energy investment interest and electricity generation cost analysis for Turkey, *Applied Energy, 87*, 2574–2580.

20. Eskin, N., Artar, H., & Tolun, S. (2008). Wind energy potential of Gökçeada Island in Turkey, *Renewable and Sustainable Energy Reviews, 12*, 839–851.

21. Jangamshetti, S. H., & Rau, V. G. (1999). Site matching of wind turbine generators: a case study, *IEEE Transactions on Energy Conversions, 14*, 1537–1543.

22. Köse, R. (2004). An evaluation of wind energy potential as a power generation source in Kütahya, Turkey, *Energy Conversion and Management, 45*, 1631–1641.

23. Lu, L. Yang, H., & Burnett, J. (2002). Investigation on wind power potential on Hong Kong islands—an analysis of wind power and wind turbine characteristics, *Renewable Energy, 27*, 1–12.

24. Mayhoub, A., & Azzam, A. (1997). A survey on the assessment of wind energy potential in Egypt, *Renewable Energy, 11*, 235–247.

25. Akpinar, E. K., & Akpinar, S. (2004). Determination of the wind energy potential for Maden-Elazig, Turkey, *Energy Conversion and Management, 45*, 2901–2914.

26. Genc, M. S. (2010). Economic analysis of large-scale wind energy conversion systems in central Anatolian Turkey, *SCIYO.COM*, p. 131.

27. Chang, T.-J., Wu, Y.-T., Hsu, H.-Y., Chu, C. R., & Liao, C.-M. (2003). Assessment of wind characteristics and wind turbine characteristics in Taiwan, *Renewable Energy, 28*, 851–871.

CHAPTER 7

EX-MINE AREA SUSTAINABILITY: EVALUATION OF LEVELS OF CONTAMINATION DUE TO HEAVY METALS FROM SOIL CHARACTERIZATION

SABARIAH ARBAI, AZINOOR AZIDA ABU BAKAR, KAMARUZZAMAN MOHAMED, and ZAINAB MOHAMED

Faculty of Civil Engineering, Universiti Teknologi MARA, Shah Alam, Malaysia

CONTENTS

OVERVIEW

Mining, one of the oldest industries in Malaysia and is once the biggest sector for the country's economy. However, this activity may lead to destruction especially to the environment due to the heavy metals accumulated in the soil that was left by previous mining activities. Some of these areas are potential for reclamation and thus require sustainability measures of environmental concern. This research that was carried out by taking soil samples from Bestari Jaya, Selangor ex-mining area would determine physical and chemical characteristic of contaminated soil. Sample taken from the site consist of four different point where marked as P1, P2, P3 and P4. The test that was done to the soil was physical test and chemical analysis. For the physical test, moisture content, particles size distribution, plastic limit and liquid limit and particle density been determined. From the test, it was found that overall sample collected are clay of intermediate plasticity with average moisture content of 21.45% and average plasticity index 12%. Particle density of soil was between 2 mg/m^3 to 4 mg/m^3. The metal traced in the soil in terms of their concentration levels by using DR 5000 spectrophotometer. In this research, contents of cadmium, zinc, copper and chromium in the soil samples were found quite high. From the chemical analysis done, total average for all heavy metal in soil were cadmium 2716.43 mg/kg, zinc 3178.57 mg/kg, copper 4000 mg/kg and chromium 2807.15 mg/kg. The result was further analyzed by referring the standard issued by Interdepartmental Committee on the Redevelopment of Contaminated Land, ICRCL and applying Single Factor Pollution Index to assess the level of contamination of the soil. The level of contamination of the soil is in range 4 to 5 and it is seriously polluted. Hence, further research should be carried out to rehabilitate the soil at the site.

7.1   INTRODUCTION

Soil is an important natural element in the world whereby as a thin layer of organic and inorganic materials it covers the surface of the earth. Soil contamination has severely increased all over the world, over the last few decades and it affects many things especially in human daily lives, concerning health and safety and as well as other living things. Soil contamination can be defined as the presence of a substance or agent in the soil as a result of human activity emitted from moving sources, from a large area or many sources and cause changes in soil natural characteristic [1]. Previous studies stated that reported major soil contamination due to heavy metal presence, are usually from abandoned mine areas, as mining is one of the most important sources of heavy metals contamination in the environment and especially to the soil [2]. *Studies were also carried out on index biotic* and the health of freshwater ecosystems which are vital to insure their safe and *sustainable* continued use.

Heavy metal can be defined as a mineral that occur naturally. Some of heavy metal is necessary for plant growth. However, excessive quantities or high concentration of heavy metal may cause adverse effects on the humans and animals health as well as food chain. According to United States Department of Agricultural (2000) through the article titled Heavy Metal Soil Contamination, it stated that heavy metals are potentially con-taminated soils at landfills sites or in around mining waste piles and tail-ings [3]. Heavy metals most frequently associated with soil contamination are such as cadmium, copper, zinc, chromium, lead and nickel. Soil con-tamination due to heavy metals differs compared to air or water pollution since heavy metals is able to accumulate and stay in soil much longer than in other compartments of the biosphere [4].

Soil contamination especially those caused by heavy metals is a seri-ous environmental problem and in this study copper, zinc, cadmium and chromium were selected heavy metals to be identified as the heavy metals presence in the soil sample collected from site. Fish and aquatic organ-isms are considered excellent bio-indicators to measure the abundance and availability of metals in the aquatic environments that may pose several severe health hazards to the aquatic consumers [5].

Soil issues related to environment especially with heavy metal pollu-tion were discussed in recent years all over the world [6]. Mining, manu-facturing, and synthetic products consumption can result in heavy metal contamination of urban and agricultural soils [3]. Malaysia history has recorded tin mine as one of the oldest industries since 1820s and the Year 2002 Malaysian mining industry report stated that mining activities has generated about RM 372.52 million to the economy [7]. However, mining activities identified as one of the major problems to the environment since and affect other living kinds as it cause serious damage to soil over a long period of time.

The concentrations of heavy metals as well as the characteristic of contaminated soil would be evaluated by physical testing and chemical analysis and to elucidate better, the levels of contamination of soil would also be assessed by applying single factor pollution index.

The study would determine the physical and chemical characteristic of collected soil samples taken at the selected ex-mine area in Bestari Jaya, Selangor, with concentration levels selected heavy metals of copper, zinc, chromium and cadmium and to classify the level of pollution in the soil.

7.2   THEORETICAL BACKGROUND

Single factor pollution index method can be used to estimate the degree of heavy metals pollution [8, 9]. This method is often used for the evaluation of the overall environment of the region soil. Besides that, this method also uses to evaluate the soil environment quality.

The formula of single factor index method can be expressed as in Equation 7.1.

$$P_i = C_i/S_i \qquad\qquad (7.1)$$

where P_i represents the environmental quality index of pollutant i; C_i is a concentration of pollutant i(mg/kg) and S_i represents the evaluation standard of heavy metal [8, 9]. Based on the index above, soil heavy metal was classified into five grades as presented in Table 7.1.

TABLE 7.1   Classification Criteria's for Polluted Index of Soil Heavy Metal Pollution

Grade/Level	Single Index	Appraisal result
0	$Pi \leq 0.7$	Safety domain
1	$0.7 < Pi \leq 1.0$	Precaution domain
2	$1.0 < Pi \leq 1.2$	Micro Polluted Domain
3	$1.2 < Pi \leq 2.0$	Slightly polluted domain
4	$2.0 < Pi \leq 3.0$	Moderately polluted domain
5	$Pi > 3.0$	Seriously polluted domain

For this research the level of contamination of soil samples of Bestari Jaya site in Selangor was then assessed and categorize according to the chemical data guidance by the Inter-department Committee on the Redevelopment of Contaminated Land (ICRCL) (1987), Guidance Note 59/83 [10]. The possible action values are as shown in Table 7.2.

TABLE 7.2 Possible Action Values

Contaminant	Value* (mg/kg)
Arsenic	69
Cadmium	15
Chromium	664
Lead	813
Mercury	10
Selenium	17
Boron	100
Copper	423
Nickel	376
Zinc	1665

7.3 MINING ACTIVITIES

Mining is the process of extraction of minerals or other geological materials from the earth, an ore body, vein or coal seam. The term also includes the removal of soil.

Material of mining includes base metals, valuable metals, iron, coal, diamonds, uranium, rock salt, oil shale, limestone and also potash. Mining is one of the most frequent method to be used all over the world to get or obtain mineral resources include heavy metal. Study carried out on water samples ex-mining lakes in Perak, Malaysia found turbid and containing slightly high concentration of lead [11].

Tin mining is one of the oldest industries in Malaysia since 1820s and have resulted in about 113,700 hectors of tin tailings that nowadays lead to many environmental problems [12]. Environmental problems include soil contamination, groundwater pollution, landscape damages as well as disturbing the natural habitats of flora and fauna.

7.4   METHODOLOGY

7.4.1   STUDY AREA

The study area is located at Bestari Jaya which is ex-mining site and is one of the catchment areas in Selangor. Based on the previous record, Bestari Jaya mining area was one of the largest tin mining areas in Malaysia. Due to mining activities, the soils at this area are highly potential to be contaminated by heavy metal. Soil assessment is needed to be done in order to identify the presence of contaminants which is heavy metal and the effect of the contaminants to the soil itself.

7.4.2   SAMPLING

The soil sample was taken at four different points around the area. The point marked as point P1, P2, P3 and P4. It was tested in laboratory in order to assess the physical and chemical characteristics. For this research, chemical analysis was interested on the concentration of heavy metal that accumulated in the soil. For that purpose, type of heavy metal that has been decided to be traced is such Cadmium (Cd), Zinc (Zn), Copper (Cu), and Chromium (Cr).

Soils samples of 3 kg were taken at depth of 1 m to 1.5 m, at every point P1, P2, P3 and P4 of the location. Undisturbed samples were extracted using standardized two-pieces aluminum pipes of each 262 mm and 200 mm respective length, sleeved as 71 mm diameter tube samplers. Each respective specimen in the tube samplers were together placed in newspaper wrapped in plastic bags and Para-film tape sealed, to prevent samples from damages as well as to retain the original condition of the soil samples. Every plastic packing containing soil samples were labeled relevant information of respective locations and sampling points using permanent markers to the accordingly recorded data. Samples were transported and assembled in the Geotechnical Laboratory of Civil Engineering Faculty, UiTM at room temperature and queued for laboratory tests.

7.4.3   MEASUREMENTS

Soils carry biological and dynamic ecosystem functions that support a diversity of life and its sustainability. This relates significantly to its physical and chemical properties. The physical tests conducted on the samples determines the soil physical characteristic and this test provides information to relate soil contamination due to heavy metals to physical characteristic of the contaminated soil. Particle size distribution from dry sieving test was carried out in order to analyze grain size of soil sample collected. Particle size distribution of a soil can determined the content of water retained, the rate of water drained and also indirectly affect the level of contaminants entrapped in the soil. The tests are particle size distribution-sieve analysis, cone penetration test, atterberg limit, moisture content and particle density of soil samples, tests were in accordance to the British Standard, BS1377:1990 [13].

Chemical analysis of the soil samples is important in determining the chemical characteristic of soil samples and test results would directly give the concentration of heavy metals presence in the soil samples. Types of heavy metals concerned in this study are cadmium, Cd, copper, Cu, zinc, Zn, and chromium, Cr. Samples were earlier oven dried and digested using *Hach Digesdahl* digestion apparatus, with combination of sulfuric acid with hydrogen peroxide at temperature 400°C to allow the process to happen. Respective tests are in accordance to the *American Public Health Association (APHA) (1995)* Standard Methods for determination of Water and Wastewater [14] and were carried out for determination of level of concentration using DR5000 spectrophotometer, by adopting porphyrin method to test for copper presence, while for chromium, zinc and cadmium, the tests were performed with dithizone method. The determination of pH was done by using pH meter.

7.5 RESULTS AND DISCUSSION

Test conducted for the data analysis includes the physical and chemical characteristics of soil samples.

7.5.1 PHYSICAL TESTS

Physical characteristic of the collected soil samples was first evaluated by its moisture content. For the moisture content, soil samples which were oven dried for about one day and weights were recorded before and after the oven drying and moisture contents are summarized in Table 7.3.

TABLE 7.3   Moisture Content of Soil Samples

Point	Moisture content, (%)
P1	19.54
P2	22.66
P3	11.64
P4	31.95

It is observed that, soil samples which were from points nearby the pond might have higher level of moisture content compared to those further away from the pond due to ground water table level differences for both locations. On the other hand, size distribution of the soil particle also contributes to the differences of moisture content of the soil samples. This is because, soils with more fine particles are considered to have higher moisture content compared to sandy soil. Clayey soil tends to hold or trap more water within its particles when compared to sandy soil which consist of loose grain particles and tend to not hold more water within its particles.

For the physical properties of the soil samples the average constituents of the soil are gravel 20.55%, sand 38.39%, silt 33.54% and clay 14.86%. Based on the proportion of the size distribution obtained, more than half of which, an amount of 58.94% of the soil samples has made up of gravel and sand while about 41.06% of soil particles made up of silt and clay.

From this test, with each point of soil samples particle density run three times to calculate for average particle density for each point, so as to indicate that particle density values of a soil sample collected may be influenced by the classification of soil that was determined by its type of soil or soil classifications. The result shows that soil with the least plasticity has higher value of particle density which is about 3.7 mg/m^3 while others shows an average value of particle density which is around 2.3 mg/m^3–2.5 mg/m^3.

7.4.2   CHEMICAL TESTS

The results of chemical analysis obtained from the chemical test with prior digestion of the soil samples carried out to get the dilute solutions and chemically analyzed as according to the respective laboratory tests. The mathematical analysis produces the results as tabulated in Tables 7.4–7.7, respectively.

TABLE 7.4 Level of Contamination for Cadmium, Cd Metal

Point	Pollutant concentration, Ci (mg/kg)	ICRCL Standard, Si	Pi	Level
P1	777.14	15	236.2	5
P2	3500.00	15	232.4	5
P3	1196.86	15	192.4	5
P4	951.43	15	63.4	5

TABLE 7.5 Level of Contamination for Zinc, Zn Metal

Point	Pollutant concentration, Ci (mg/kg)	ICRCL Standard, Si	Pi	Level
P1	3542.86	1665	2.1	4
P2	3485.71	1665	2.1	4
P3	2885.71	1665	1.7	3
P4	2800.00	1665	1.7	3

TABLE 7.6 Level of Contamination for Copper, Cu Metal

Point	Pollutant concentration, Ci (mg/kg)	ICRCL Standard, Si	Pi	Level
P1	3942.86	423	9.3	5
P2	4400.00	423	10.4	5
P3	3628.57	423	8.6	5
P4	4028.57	423	9.5	5

TABLE 7.7 Level of Contamination for Chromium, Cr Metal

Point	Pollutant concentration, Ci (mg/kg)	ICRCL Standard, Si	Pi	Level
P1	2428.57	664	3.7	5
P2	2942.86	664	4.4	5
P3	3114.29	664	4.7	5
P4	2742.86	664	4.1	5

The results shows that among all the type of heavy metal considered in this research, Zn contained the least amount that accumulated in the soil. The other type of heavy metals such as Cd, Cr and Cu show quite high level accumulated in the soil.

Figure 7.1 show the distribution of heavy metal at sample locations P1, P2, P3 and P4 for the area of Bestari Jaya especially around catchment area where the soil sample has been taken. The distribution of heavy metal showed in terms of its level of pollution that has been calculated by applying the single factor pollution index formula as in Eq. (7.1) and referring to ICRCL standard.

Square point at each point of soil sample collected that are marked with a, b, c and d represents the level of contamination for Cadmium, Zinc, Copper and Chromium, respectively.

Only Zinc metal has varies in level of contamination at each point whereby points P1 and P2 are at 4th level of contamination and points P3 and P4, are at 3rd level. While, the other type of heavy metal at all points

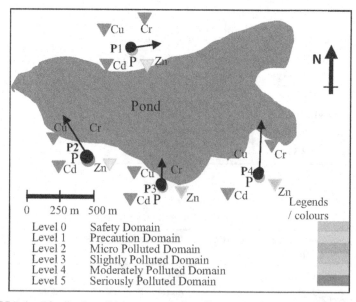

FIGURE 7.1 Distribution of heavy metals in soil samples for Bestari Jaya relocations P1, P2, P3 and P4.

were identified at 5th levels of contamination and which were indicated seriously polluted.

7.6   CONCLUSION AND RECOMMENDATION

This research has found that the types of soil at the area chosen are made up of Clay soil through the physical tests that have been carried out. The chemical analysis finding from this research is alarming. The results of heavy metal content in a soil sample collected show the concentration are quite high. Heavy metal concentration have been further evaluated through Single Factor Pollution Index and found that overall level of contamination for all the type of heavy metal which are Cadmium, Copper and Chromium was at the fifth level which the level is seriously polluted. Only Zinc metal show slightly varies where the concentration at P1 and P2 are consider moderately polluted which is at 4th level and at P3 and P4 are slightly polluted where at the 3rd level. After been compared to the other different parts of study area, it is concluded that Bestari Jaya soil has high risk in pollution. This may become worst when all the water flow through to the catchment area and get polluted too.

Therefore, for comprehensive sustainability measures, an extensive research should be carried out for this area especially regarding heavy metal contamination. The research should cover not only soil specimens but also the nearest surrounding aquatic environment [15]. This would enable to assess the potential health risks associated for the possibility of fish consumption from these catchments. Some recommendations regarding this research that can be taken into consideration for future action are:

1. The research should cover larger area by taking many points soil samples and as well as aquatic and plants specimens collected so that the results obtained will have better representation.
2. The types of heavy metal concern in the research can be in a various type rather than just concerning on cadmium, Zinc, Copper, and Chromium, other metals such as Arsenic, As; Plumbum, Pb; Cobalt, Co and many more should also be considered since they are also very dangerous to the environment.

3. Water samples from the nearest water bodies or catchment area should also be tested to determine whether heavy metals in soil might have any influence to the water quality.
4. Other methods can also be used such as by using X-Ray Diffraction (XRD) testing method so that the results obtained from the physical and chemical laboratory tests can be compared to the result from the other method and the accuracy can be ensured.
5. More assessment studies of heavy metals pollution in ex-tin mine catchments should be performed in order to provide more useful data and help to ensure the quality of fish and safeguard of human health.
6. Future research should provide rehabilitation and reclamation action that is to be taken so that the research is not just to identify the pollution occurred but also action to overcome the pollution matter.

ACKNOWLEDGEMENT

The authors would like to express their deepest appreciation to Universiti Teknologi MARA through the Research Management Institute (RMI) and Ministry of Higher Education (MOHE) for the FRGS (600-RMI/ST/FRGS5/3/Fst (278/2010) for the financial support to this research.

KEYWORDS

- ex-mine
- heavy metals
- land reclamation
- soil characterization
- soil classification
- soil contamination

REFERENCES

1. Panos Panagos, Marc Van Liedekerke, Yusuf Yigini, & Luca Montanarella. (2013). "Contaminated Sites in Europe: Review of the Current Situation Based on Data

Collected through a European Network." *Journal of Environmental and Public Health*, Article ID 158764, 11 pp.

2. Alshaebi, F. Y., WanYaacob, W. Z., Samsudin, A. R., & Alsabahi, E. (2009). "Risk Assessment at Abandoned Tin Mine in Sungai Lembing, Pahang, Malaysia." *Electronic Journal of Geotechnical Engineering (EJGE)*, *14*.

3. United States Department of Agriculture (Natural Resources Conservation Service), USDA "Effects of Land Management Quality-Heavy Metal Soil Contamination", (2000). *Soil Quality – Urban Technical Note No. 3.*

4. Haribabu, T. E., & Sudha, P. N. (2011). "Effect of Heavy Metals Copper and Cadmium Exposure onthe Antioxidant Properties of the Plant Cleome Gynandra." *International Journal of Plant, Animal and Environmental Sciences*, *1*(2).

5. Ahmad, A. K., Sarah, A., & Al-Mahaqeri (2015). "Human Health Risk Assessment of Heavy Metals in Fish Species Collected from Catchments of Former Tin Mining", *International Journal of Research Studies in Science, Engineering and Technology*, *2*(4), 9–21.

6. Liao, G. L., Liao, D. X., & Li, Q. M. (2008). "Heavy Metals Contamination Characteristic in Soil of Different Mining Activity Zones." *Transaction of Nonferrous Metals Society in China*, *18*, 207–211.

7. Ashraf, M. A., Maah, M. J., & Yussof, I., (2011). "Sand mining effects, causes and concerns: A case study from Bestari Jaya, Selangor, Peninsular Malaysia." *Scientific Research and Essays*, *6*(6), 1216–1231.

8. Yang L., Wu S. "Assessment of Soil Heavy Metal Cu, Zn, and Cd pollution in Beijing, China." *Advanced Materials Research*, *(2012)*. Vols. 356–360, pp 730–733.

9. Zhao R., Guo W., Sun W., Xue S., Gao B., & Sun W. (2012). "Distribution Characteristic and Assessment of Soil Heavy Metal Pollution around Baotou Tailing in Inner Mongolia, China." *Advanced Materials Research*. *356–360*, 2730–2736.

10. Guidance Note 59/83, *Guidance on the Assessment and Redevelopment of Contaminated Land, Department of the Environment, London.* Interdepartmental Committee on the Redevelopment of Contaminated Land, (ICRCL), (1987).

11. Kalu Uka Orgi., Nasiman, B. S., Khamaruzaman, W. Y., Asadpour, R., & Emmanuel, O. (2014). "Water Quality Assessment of Ex-Mining Lakes in Perak, Malaysia as Alternative Source of Water Supply." *Applied Mechanics and Materials*, *567*(1), 177–118.

12. Muhammad Aqeel, A., Mohd. Jamil, M., & Ismail, Y. (2010). "Study of Water Quality and Heavy Metals in Soil & Water of Ex-Mining Area, Bestari Jaya, Peninsular Malaysia." *International Journal of Basic & Applied Sciences IJBAS-IJENS*, *10*(3).

13. British Standards BS 1377–2:1990. Methods of test for Soils for Civil Engineering Purposes Classification Tests.

14. American Public Health Association (APHA). *Standard Methods for the Examination of Water and Wastewater.* (1995). ISBN 0-87553-207-1.

15. Meng Zhao-Hong, Zheng Yuan-Fu, & Xiao Hai-Feng (2011). "Distribution and Ecological Risk Assessment of Heavy Metal Elements in Soil." *Advanced Materials Research*, *183–185*, 82–87.

CHAPTER 8

DEVELOPMENT OF NEW FORMULATION FOR AUTOMOTIVE WEATHERSTRIP SEAL

FATIMAH ZUBER, ENGKU ZAHARAH ENGKU ZAWAWI, and DZARAINI KAMARUN

Faculty of Applied Sciences, Universiti Teknologi MARA, Shah Alam, Malaysia

CONTENTS

OVERVIEW

Development of low cost products with similar or better properties is vital to its sustainability in the competitive global market. This research embarks on compositional analysis of the hard part of a weatherstrip seal product that surpassed customer's specification and development of a new formulation which incorporated ground calcium carbonate (GCC) as a secondary filler for cost reduction. Qualitative and quantitative analysis of the conformable rubber product were carried out using soxhlet extraction for extraction of the additives; and characterization techniques which include spectroscopic, chromatographic, X-ray diffraction and several thermal techniques. Test pieces compounded from 6 new formulations were tested for their mechanical performance. Thermogravimetric analysis (TGA) and differential scanning calorimetry (DSC) confirmed the presence of EPDM and carbon black in the compound. Fourier Transform Infrared (FTIR) and chromatographic analysis confirmed the processing oil in the compound to be of paraffinic origin. Presence of talc filler was proven by elemental analysis using X-ray Diffraction (XRD). Addition of GCC tends to reduce the tensile strength, tear strength and elongation at break as expected. However, the reduction was not high enough that could lead to the failure of product. Overall, the mechanical properties of all the new formulated vulcanizates varied within customer's specification.

8.1   INTRODUCTION

Weatherstrip seal functions to prevent leakage of liquid such as water into the car. Weatherstrip seal for cars produced from co-extrusion of hard and sponge rubber where the hard part contributed about 86% of the total volume. Company TY is a manufacturer of the weatherstrip used in this study. The compounded material used is of unknown materials and composition; and is claimed to be of high cost. It is therefore the interest of the company's administration to identify the composition of the compound in an effort to prepare new compounds of competitive performances. The hard part of the weatherstrip seal was selected for analysis due to its higher volume contribution. Reverse engineering was adopted for this purpose. Reverse engineering has been practiced in the past few decades as a process of extracting knowledge from anything man-made artifact. It is a most well-known application for the purpose of developing competing products and help designers to obtain the required information of the benchmark product in a short time.

Elastomers were often used for sealing applications that require flexibility. Ethylene propylene diene monomer (EPDM) is the largest non-tire elastomer typically used in automotive seals. EPDM market is still growing. Many extruded products for automotive industry are now made of this rubber including window seals, radiator hoses and many other profiles. EPDM rubber is also known to accept high filler loading due to its low density. However, high content of raw EPDM contributed to high production cost; thereby forcing rubber products manufacturers to modify existing formulations to meet customer requirements for 'within specification' products at competitive price. The overall amount of EPDM consumption could be reduced by introducing cheapening fillers into the compound.

Fillers have been widely used to cheapen the cost while maintaining or improving the mechanical properties of the compound. Fillers are grouped into reinforcing filler, one which enhance the mechanical properties including strength and stiffness of rubber; and non-reinforcing fillers. Non-reinforcing fillers are usually applied as diluents or extenders to reduce cost. Carbon black (CB) is conventional reinforcing filler and is the major constituent of industrial rubber products. Among the many grades available, fast extruding furnace (FEF) is one of the most widely used in the rubber industry. Calcium carbonates of the natural calcite–group minerals in the form of ground calcium carbonate (GCC) was used in the formulation of the new product as it is well-known as non-reinforcing filler and is abundant in nature as limestone. GCC are widely used in the industry since it is relatively inexpensive. Much work has been reported on the usage of calcium carbonate as fillers for rubber compounds [1–3] but none reports high loading of calcium carbonate as secondary filler.

This paper described the analytical techniques used to determine the components present in the hard part of a weatherstrip seal and its composition using various chemical and instrumental techniques. Soxhlet extraction was conducted to extract additives present in the compound which were then analyzed using gas chromatography – mass spectrometry (GC-MS). Thermogravimetric analyzer (TGA), differential scanning calorimetry (DSC) and X-ray diffraction (XRD) were used to identify the insoluble components. Information extracted from the analysis was used to reformulate several formulations of a new compounding material incorporating GCC for cost reduction. GCC was varied from 40 to 140 phr as secondary filler. The purpose is to study the extent of incorporation of GCC affect the properties of the compound. The mechanical properties were determined to identify their performance in accordance to product specification which is justified as follows: Hardness value: 80–90 IRHD; Tensile strength: $\geq$7.0 MPa; Elongation at break: $\geq$200% (minimum); Tear strength: $\geq$147.0 N/mm.

8.2 EXPERIMENTAL

8.2.1 CHEMICALS AND MATERIALS

The hard part of weatherstrip seal (Sample W), raw EPDM (sample E) and paraffin oil was supplied by Company TY. Acetone is supplied by Chem AR.

8.2.2 EXTRACTION OF PROCESSING OIL

The extraction of processing oil from Sample W was carried out in a soxhlet extractor using acetone as solvent. Solvents from extracts were removed using rotary evaporator and liquid obtained were washed with acetone and dried in a vacuum oven at room temperature for 24 hours.

8.2.3 THERMAL ANALYSIS

Samples E and W were analyzed using thermogravimetric (TG) analyzer in two steps: Initial heating from 50 to 500°C at a heating rate of 20°C/min under nitrogen atmosphere followed by oxygen gas environment up to 800°C under the same heating rate. Insolubles from the soxhlet extraction which remained in the thimble after extraction were analyzed at temperatures 50 to 500°C. DSC analysis was performed under nitrogen atmosphere at a heating rate of 20 ml/min. Samples were initially cooled from ambient temperature to –80°C and then heated to 0°C at a rate of 10°C/min.

8.2.4   FOURIER TRANSFORM INFRARED (FTIR) ANALYSIS

Spectrum One Fourier Transform Infrared (FTIR) spectroscopy from Perkin Elmer was used with attenuated total reflectance (ATR) sampling technique. Tests were conducted over 16 scans in the range of 4000–600 cm^{-1} at a resolution of 4 cm^{-1}.

8.2.5   CHROMATOGRAPHIC ANALYSIS

The extracted oil was analyzed using a gas chromatography-mass spectrometer (Agilent Technologies 6890N GC) to determine its composition. The chromatographic column for the analysis was a 5% phenyl methyl polysiloxane capillary column (25 mm × 30 mm × 0.25 μm). The carrier gas used was helium at a flow rate of 1.0 ml/min. The sample was analyzed with the column heated from 50°C to 300°C at a heating rate of 5°C/min.

8.2.6   X-RAY DIFFRACTION (XRD) ANALYSIS

Sample W was analyzed for presence of inorganic fillers using XRD. The test was carried on a PAN analytical XRD analyzer using Cu K$_\alpha$ radiation over 2θ range of 5 to 30°. The data were analyzed and re-plotted using X'pert Highscore Plus software.

8.2.7   FORMULATION AND COMPOUNDING OF NEW EPDM VULCANIZATES

Results obtained from compositional analysis of sample W were used to formulate six new formulations of vulcanizates incorporating GCC as secondary filler. Compounding of the vulcanizates was carried out on a conventional laboratory-sized two-roll mill machine according to ASTM D 3182. Formulations of the compounding ingredients expressed as parts per hundred rubber (phr) for seven different GCC loadings

(20–140) phr labeled as A, B, C, D, E, F and G are shown in Table 8.1. Fixed amount of other typical curatives of a rubber vulcanizates were added to the formulations.

TABLE 8.1 Formulations of New Rubber Vulcanizates with Various GCC Loadings

Ingredient	Compound (phr)						
	A	**B**	**C**	**D**	**E**	**F**	**G**
EPDM	100	100	100	100	100	100	100
Paraffin Oil	40	40	40	40	40	40	40
CB	100	100	100	100	100	100	100
GCC	20	40	60	80	100	120	140
Zinc oxide	10.0	10.0	10.0	10.0	10.0	10.0	10.0
Stearic acid	1.5	1.5	1.5	1.5	1.5	1.5	1.5
Antioxidant	1.0	1.0	1.0	1.0	1.0	1.0	1.0
CBS	0.3	0.3	0.3	0.3	0.3	0.3	0.3
MBTS	0.3	0.3	0.3	0.3	0.3	0.3	0.3
ZDBC	0.2	0.2	0.2	0.2	0.2	0.2	0.2
TMTD	0.2	0.2	0.2	0.2	0.2	0.2	0.2
Sulphur	0.7	0.7	0.7	0.7	0.7	0.7	0.7

8.2.8 CURE CHARACTERISTIC OF EPDM VULCANIZATES

The cure characteristics were measured using oscillating disc Mosanto rheometer 100 at a temperature of 180°C. The process was performed for 6 minutes to obtain the optimum cure time (t_{90}) and cure rate index (CRI). T_{c90} corresponds to the time required to achieve 90% of the maximum torque (M_{90}), which was determined by the following Equation 8.1 and CRI was determined using Equation 8.2.

$$M_{90} = M_L + 0.90 \, (M_H - M_L) \tag{8.1}$$

$$CRI \, (\% \, min^{-1}) = \frac{100}{t_{c90} - t_{s2}} \tag{8.2}$$

where M_L, M_H and t_{s2} are minimum torque and maximum torque and scorch time, respectively.

8.2.9   MECHANICAL PROPERTIES OF EPDM VULCANIZATES

Each vulcanizate were compression molded at 180°C using cure time, t_{c90} determined. Tensile test was performed according to ISO 37 on dumbbell-shaped tensile test piece type II (length: 75.0 ± 1.0 mm; width: 4.0 ± 0.1 mm) which was cut from 2.0 mm thick molded sheet. Tear test was carried out according to ASTM D624 on angle cut tear test piece. The test pieces (length: 102.0 ± 0.5 mm; width: 19.0 ± 0.05 mm) were prepared using die C cutter stamped on 2.0 mm thick molded sheet. The measurements of both tensile and tear test were carried out at room temperature on Instron 5569 tensile tester with gauge speed of 500 mm/min. Hardness test was also conducted using Wallace Cogenix H14 dead load hardness tester according to ISO 48.

8.2.10   MORPHOLOGICAL STUDY OF EPDM VULCANIZATES

Morphological studies of the EPDM vulcanizates were carried out using a Field Emission Scanning Electron Microscope (FESEM) SUPRA–40 VP with software INCA Suite versions 4.12 with the magnification, 5,000× and it was connected to Energy Dispersive X-ray analyzer (EDX). Cross section of the sample was cryo-fractured in liquid nitrogen to avoid any possibility of phase deformation during the fracture and then coated with a layer of gold to eliminate electrostatic charge build-up during examination.

8.3   RESULTS AND DISCUSSION

8.3.1   THERMAL ANALYSIS

Figure 8.1 shows the TG curves of sample W and sample E. Sample E which was raw EPDM decomposed at 478.9°C with a sharp decomposition profile. Sample W shows two decomposition temperatures, T_d at 460.5°C and 648.3°C which corresponded to the degradation of rubber component and carbon filler, respectively. The first decomposition profile of the rubber component was not sharp but decreased gradually from 250 to 500°C

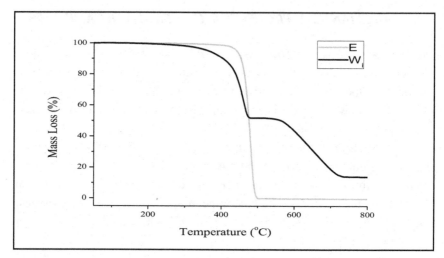

FIGURE 8.1 TG curve of raw EPDM (E) and the hard part before extraction (W$_i$).

indicating the presence of additive(s) which decomposed at temperatures lower than the rubber component. Markovic et al. [4] identified the decomposition of EPDM to be in the range of 450 to 490°C. Decomposition of carbon black occurred in the range of 550°C to 650°C [5]. Carbon black of fast extruding furnace (N550) was known to decompose at 648°C [5]. The amount of rubber components (plus additive) and carbon black in W was determined to be 50 and 35 wt%, respectively as calculated from the mass loss upon degradation. A final unburnt residue of 15 wt% remained. The residue was identified to be talc as described in section 3.4.

Figure 8.2 showed a TG curve of sample W before (W$_i$) and after extraction (W$_f$). The amount of oil present was determined to be 15 wt% based on the difference in mass % before and after extraction. Consequently, the amount of the rubber component was determined to be 35 wt%. Identity of the processing oil was determined and discussed in Section 8.3.2 and 8.3.3.

DSC analysis confirmed further the identity of the rubber component to be EPDM. The glass-transition temperature (T$_g$) of sample W before extraction (W$_i$) and after extraction (W$_f$) (see Figure 8.3) were −58°C and −47°C, respectively. EPDM rubber was known to have a T$_g$ of −55°C

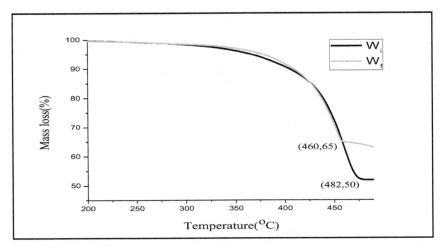

FIGURE 8.2   TG curve of hard part before extraction (W$_i$) and after extraction (W$_f$).

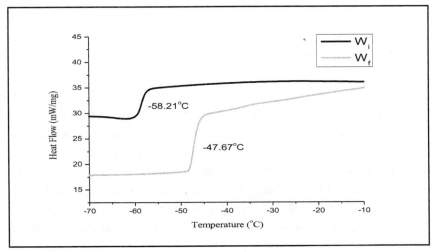

FIGURE 8.3   DSC curve of the hard part before extraction (W$_i$) and after extraction (W$_f$).

[5]. The processing oil in W$_i$ acts as a plasticizer which caused the T$_g$ to be lower than W$_f$. Pradhan et al. reported that addition of plasticizer helps increased the local chain flexibility hence lowering the T$_g$ [6].

8.3.2   FTIR ANALYSIS

Figure 8.4 compared the FTIR spectra of the oil extracted from W (denoted by T oil) and standard paraffinic oil (denoted by P oil).

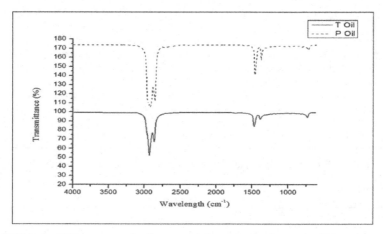

FIGURE 8.4   FTIR spectra of oil extracted from the hard part (T Oil) and standard paraffinic oil (P oil).

T oil showed absorption bands at 2920 cm^{-1}, 2854 cm^{-1}, 1464 cm^{-1}, 1379 cm^{-1} and 726 cm^{-1}. The bands are similar to those of P oil which are at 2924 cm^{-1}, 2861 cm^{-1}, 1465 cm^{-1}, 1375 cm^{-1} and 725cm^{-1}. The bands in the region of 2850–3000 cm^{-1} are attributed to stretching of CH_3 and CH_2 bonds of aliphatic hydrocarbons. CH bending and rocking are noted in the regions of 1370–1460cm^{-1} and at 723cm^{-1}, respectively. Pistor et al. also reported similar bands of paraffinic oil extracted from EPDM residue [7]. Thus, the oil analyzed in this study was of paraffinic oil since it contained only saturated hydrocarbons.

8.3.3   CHROMATOGRAPHIC ANALYSIS

Chromatographic analysis using GC-MS was conducted to identify components of oil extracted. Table 8.2 and showed the components identified (spectrum not shown) of oil extracted from W. It is concluded that the oil extracted is of paraffinic origin since it contained only alkane components.

TABLE 8.2 Component Identified From GC/MS Analysis of Oil Extracted From W

Component	Retention time, s	Area, %
Heneicosane	32.78	2.32
Docosane	34.42	4.56
Heptadecane	36.13	10.39
Tetracosane	37.77	16.06
Eicosane	39.35	17.34
Heptadecane	40.87	18.32
Heptacosane	42.34	13.62
Octadecane	43.75	8.69
Hexadecane	45.12	5.52
Triacontane	46.44	3.18

8.3.4 X-RAY DIFFRACTION (XRD) ANALYSIS

The residue remained from TGA analysis was determined using XRD analysis. Figure 8.5 showed the XRD patterns for W and standard talc. Presence of talc in W was confirmed from peaks observed at 9.2°, 19° and 28.8° Researches done by Pervin et al. [8] and Benito et al. [9] showed similar XRD patterns of talc.

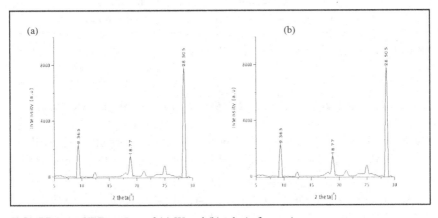

FIGURE 8.5   XRD pattern of (a) W and (b) talc (reference).

8.3.5 CURE CHARACTERISTICS OF EPDM VULCANIZATES

Figure 8.6 shows cure time and cure rate index of various GCC loading. Both curves showed no significance difference of the values. Introduction of 20 phr GCC showed t_{c90} and CRI of 2.5 min and 50.0 min^{-1}, respectively. As GCC was loaded up to 140 phr, the cure time increased to 2.8 minutes and CRI decreased to 48.8 min^{-1}. All vulcanizates showed short cure time and high cure rate index due to the use of very fast reacting ultra-accelerator, TMTD and ZBDC [10].

Basically, since GCC possesses a large particle size and poor binding interaction with rubber, its low surface area has little effect on the curing characteristics. A previous study also shows that even precipitated calcium carbonate which has smaller particle size than GCC has little effect on the curing characteristics due to its low surface area hence less binding interaction with rubber particles [2, 11].0

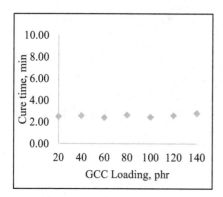 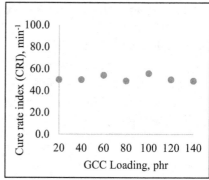

FIGURE 8.6   Cure Time and Cure Rate Index of Various GCC Loading.

8.3.6 MECHANICAL PROPERTIES OF EPDM VULCANIZATES

Figure 8.7 shows tensile strength, tear strength and elongation at break of EPDM vulcanizates at various GCC loadings. All properties decreased as GCC was loaded from 20 phr to 140 phr.

The reduction in tensile and tear strength were attributed to the non-reinforcing nature of GCC. GCC have large particle size and polar surface which does not interact well with the rubber matrix [12–14]. Increased in

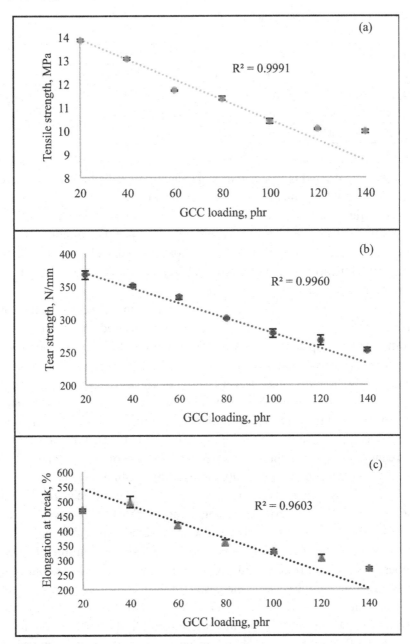

FIGURE 8.7   (a) Tensile strength, (b) tear strength and (c) elongation at break of EPDM vulcanizates with various GCC loading.

TABLE 8.3   Reduction (%) of Mechanical Properties with Addition of 40 phr GCC to EPDM Vulcanizate

Properties	Reduction (%)	
	(60–100) phr GCC	(100–140) phr GCC
Tensile strength	11.1	4.8
Tear strength	16.5	9.0
Elongation at break	21.2	18.0

GCC loading diluted the elastomer which led to the reduction in the elastic behavior of rubbers thus reducing the elongation at break. The failure in all properties was also attributed to the reduced amount (in wt %) of the reinforcing filler – CB and EPDM rubber [14, 15] as the amount of GCC added increased. Poh et al. [12] also reported similar observation when using calcium carbonate in elastomer; although a different type of elastomer and calcium carbonate were used.

As shown in Figure 8.7, a linear relationship between tensile/tear strength/elongation at break and amount of GCC loaded was observed for addition of GCC up to 100 phr. Above 100 phr GCC loading, the plot deviated from linearity with minimal reduction in the properties measured. At 100 phr GCC loading, the amount (wt %) of EPDM, CB and GCC were the same, i.e., 28.2% each. Further addition of GCC made GCC dominant over the other ingredients and the reduction in properties became minimal (see Table 8.3) due to the non-reinforcing nature of GCC. Table 8.3 shows the reduction (%) of the properties with the addition of every 40 phr GCC to EPDM vulcanizate in 60 to 100 phr and 100 to 140 phr GCC.

Figure 8.8(a) shows the hardness value of EPDM vulcanizates with various GCC loading. An increasing trend of the hardness value was obtained for all vulcanizates as GCC was increased up to 140 phr. Nair and co-workers reported a hardness value of 59 IRHD for unfilled EPDM rubber [15]. Vulcanizate A (20 phr GCC) showed a hardness value of 77 IRHD. The hardness value increased markedly due to the presence of high loading of CB filler (100 phr). When carbon black is loaded into rubbers, the stiffness of rubber increased markedly [16] due to the reinforcement effect on EPDM rubber. The hardness of vulcanizates increased further with the addition of GCC fillers up to 140 phr. The highest hardness obtained was 89 IRHD. The expected increasing trend was attributed to the fact that as more GCC fillers were incorporated into the rubber matrix, the compounds became more rigid due to the reduction of rubber chain elasticity as the filler restricted the mobility and deformability of the rubber. This argument was agreed by several other authors [17, 18].

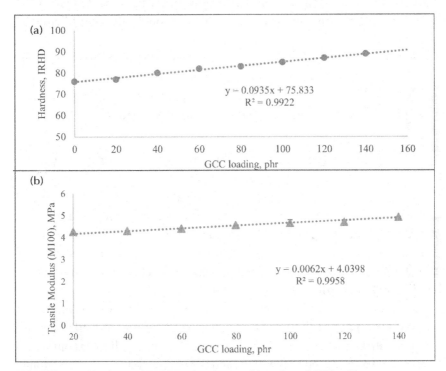

FIGURE 8.8 Hardness of EPDM vulcanizates with various GCC loading.

The range of hardness value allowed for weatherstrip application is 80 to 90 IRHD. Extrapolation of the fitted linear graph allowed one to predict the maximum amount of GCC that could be loaded while maintaining the specified maximum hardness of 90 IRHD. The maximum allowed GCC is 151.5 phr as measured from Figure 8.8 (a).

Figure 8.8 (b) shows the tensile modulus of stress at 100% strain (M100) of EPDM vulcanizates containing various GCC loadings. M100 showed a value of to 4.2 MPa when 20 phr GCC was incorporated. Further loading of GCC (up to 140 phr) increased M100 to 4.9 MPa. Based on previous study on CB fillers, tensile modulus increased along with hardness as filler loading increased due to the reinforcing effect of CB filler [12, 19]. Studies by other researchers showed that addition of 50 phr carbon black increased M100 by 4 to 5.0 MPa compared to blank sample [12, 20]. Using the linear equation generated in Figure 8.8 (b), addition of 50 phr GCC increased the M100 by 0.27 MPa. The increment was not as significant as when CB filler was utilized due to the fact that GCC is non-reinforcing filler compared to CB black.

Addition of GCC reduced the tensile strength, tear strength and elongation at break as expected. However, the reduction was minimal and did not lead to product failure. The mechanical properties for all vulcanizates conformed to product specification as required by customer.

8.3.7 MORPHOLOGICAL STUDY OF EPDM VULCANIZATES

FESEM images of cross-section of vulcanizates B and G were taken at a magnification of 5,000 times. The images were further analyzed by EDX analyzer coupled to the FESEM instrument. The different light contrast of the images reflected different morphological properties on the surface of the samples. EDX mapping on images were carried out to observe elemental distribution of fillers in the EPDM matrix.

Figure 8.9 shows EDX mapping on FESEM images of vulcanizate B (40 phr GCC) and G (140 phr GCC).Uneven distribution of fillers in the rubber matrix of vulcanizate B and G were observed. A lighter contrast of the image was assigned to the filler; and the darker contrast to the continuous phase of rubber matrix.

GCC filler was identified by the presence of calcium and higher percentages of oxygen atoms. Uneven distribution of the elements Ca, O and C was observed. Vulcanizate B showed some agglomeration of GCC as indicated by the red circle. Vulcanizate G which has higher amount of GCC showed a higher degree of uneven distribution of fillers. The distribution density of carbon atoms was lower in vulcanizate G. On the other hand, the distribution density of oxygen and calcium showed an increase due to high wt% of GCC incorporated. Accumulation of oxygen atoms in vulcanizate G was due to agglomeration of GCC as indicated by the yellow circle. Accumulation of carbon atoms at certain areas

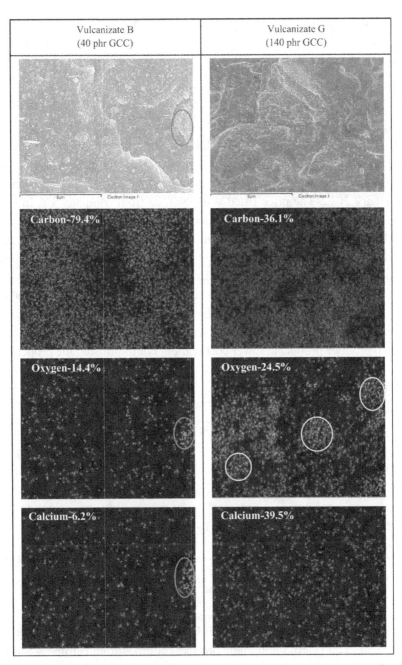

FIGURE 8.9   EDX mapping on FESEM image of vulcanizate B and G at magnification 5,000×.

indicated the possibility of agglomeration of CB fillers as well. Volume occupied by all fillers exceeded the volume available in the EPDM matrix which led to agglomeration.

8.4   MATERIAL COSTING

Table 8.4 showed a summary of material costing for vulcanizate B and G. Estimation on material cost of the vulcanizates for the production of one kilogram compounded product were calculated based on materials price as of January 2012. The price listed was obtained from company TY. In general the total cost per kg of compounded product was reduced as the amount of EPDM and CB decreased with increased in GCC loadings. The total cost for producing compounded product for vulcanizate B is USD 2.51/kg. Vulcanizate G with 140 phr of GCC incorporated cost less by USD 2.06/kg with a reduction of 17.9%. The cost was reduced abruptly due to the reduction of the main material, EPDM which contributed to the highest overall material cost. The newly formulated vulcanizates satisfied the main objectives to produce product at lower cost without compromising the properties.

TABLE 8.4   Material Costing of Vulcanizate B and G

Ingredient	Price (USD/kg)	B			G		
		phr	wt %	Cost, USD	phr	wt %	Cost, USD
EPDM	4.88	100	35	488.00	100	26	488.00
Paraffin Oil	1.68	40	15	67.20	40	11	67.20
CB	1.52	100	35	152.00	100	26	152.00
GCC	0.76	40	15	30.40	140	37	106.40
Total		280	100	737.60	380	100	813.60
Compound cost (USD/kg)		2.51			2.06		

8.5   CONCLUSION

Reverse engineering is a very useful technique for identifying product composition and formulating new products composition. Hardness served as a quality control that determined the limit of using large amount of GCC as secondary filler in this weatherstrip seal application. Compound formulations that conformed to the industry's need were successfully developed which could help manufacturers in formulating new products with comparable if not better properties at lower costs.

ACKNOWLEDGEMENT

The authors are grateful to Faculty of Applied Sciences of Universiti Teknologi MARA Shah Alam for providing the facilities to carry out the research; and Company TY for providing the research internship.

KEYWORDS

- **EPDM**
- **fillers**
- **ground calcium carbonate**
- **mechanical properties**
- **vulcanizates**
- **weatherstrip seal**

REFERENCES

1. Moonchai, D., Moryadee, N., & Poosodsang, N., (2012). Comparative Properties of Natural Rubber Vulcanizates Filled with Defatted Rice Bran, Clay and Calcium Carbonate. *Maejo International Journal of Science and Technology*, *6*(2), 249–258.
2. Chuayjuljit, S., Imvittaya, A., Na-Ranong, N., & Potiyaraj, P., (2002). Effects of Particle Size and Amount of Carbon Black and Calcium Carbonate on Curing Characteristics and Dynamic Mechanical Properties of Natural Rubber. *Journal of Metals, Materials and Minerals*, *12*(1), 51–57.
3. El-Wakil, A. A., El-Megeed, A. A. A. (2011). Effect of Calcium Carbonate, Sillitin N85 and Carbon Black Fillers on the Mechanical and Electrical Properties of The EPDM. *ARPN Journal of Engineering & Applied Sciences*, *6*(5), 24–29.
4. Markovic, G.-M., Choudhury, N. R., Dimopoulos, M., Williams, D. R. G., & Matisons, J., (1998). Characterization of Elastomer Compounds by Thermal Analysis. *Thermochimica Acta*, *316*(1), 87–95.
5. Naranjo, A., (2008). Plastics Testing and Characterization: Industrial Applications. Hanser Gardner Publications.

6. Pradhan, D. K., Choudhary, R. N. P., Samantaray, B. K., Karan, N. K., & Katiyar, R. S., (2007). Effect of plasticizer on structural and electrical properties of polymer nanocomposite electrolytes. *Int J Electrochem Sci, 2*, 861–871.

7. Pistor, V., Ornaghi, F. G., Fiorio, R., & Zattera, A. J., (2010). Thermal characterization of oil extracted from ethylene–propylene–diene terpolymer residues (EPDM-r). *Thermochimica Acta, 510*(1–2), 93–96.

8. Parvin, N., Ullah, M. S., Mina, M. F., & Gafur, M. A., (2013). Structures and mechanical properties of talc and carbon black reinforced high density polyethylene composites: Effects of organic and inorganic fillers. *Journal of Bangladesh Academy of Sciences, 37*(1), 11–20.

9. Benito, J. M., Turrillas, X., Cuello, G. J., Aza, S. D., & Rodriguez, M. A., (2011). Cordierite synthesis. A Time-resolved Neutron Diffraction Study. *Journal of the European Ceramic Society, 32*.

10. Dijkhuis, K. A. J., Noordermeer, J. W. M., & Dierkes, W. K. (2009). The relationship between crosslink system, network structure and material properties of carbon black reinforced EPDM. *European Polymer Journal, 45*(11), 3302–3312.

11. Roberts, A. D., Natural rubber science and technology. 1988: Oxford University Press.

12. Poh, B. T., Ismail, H., & Tan, K. S. (2002). Effect of filler loading on tensile and tear properties of SMR L/ENR 25 and SMR L/SBR blends cured via a semi-efficient vulcanization system. *Polymer Testing, 21*(7), 801–806.

13. Ismail, H., Zulkepli, N. N., Wei, W. H., & Nasir, M. R. J., (2013). The Influence of CB, Silica and $CaCO_3$ on Tensile and Morphological Properties of vPE/rPE/EPDM Blends. *Advanced Materials Research, 844*, 338–341.

14. Muniandy, K., Ismail, H., & Othman, N. (2012). Effects of Partial Replacement of Rattan Powder by Commercial Fillers On the Properties of Natural Rubber Composites. *BioResources, 7*(4), 4640–4657.

15. Nair, T. M., Kumaran M. G., & Unnikrishnan G., (2004). Mechanical and aging properties of cross-linked ethylene propylene diene rubber/styrene butadiene rubber blends. *Journal of Applied Polymer Science, 93*(6), 2606–2621.

16. Samsuri, A., (2013). Theory and Mechanisms of Filler Reinforcement in Natural Rubber, in Natural Rubber Material; Volume 2: Composites and Nanocomposites, S. Thomas, et al., Ed. Royal Society of Chemistry: Cambridge, UK.

17. Aguele, O. F., & Madufor, C. I., (2012). Effects of Carbonized Coir on Physical Properties of Natural Rubber Composites. *American Journal of Polymer Science, 2*(3), 28–34.

18. Egwaikhide, P. A., Akporhonor E. E., & Okieimen F. E., (2007). Effect of coconut fiber filler on the cure characteristics physico–mechanical and swelling properties of natural rubber vulcanisates. *International Journal of Physical Sciences, 2*(2), 039–046.

19. Rattanasom, N., Prasertsri S., & Ruangritnumchai, T. (2009). Comparison of the mechanical properties at similar hardness level of natural rubber filled with various reinforcing-fillers. *Polymer Testing, 28*(1), 8–12.

20. Du, A., Zhang Z., & Wu M., (2009). The effect of Pyrolytic Carbon Black Prepared from Junked Tires on the Properties of Ethylene-Propylene-Diene Copolymers (EPDM). *Express Polymer Letters, 3*(5), 295–301.

CHAPTER 9

LEATHER AND SYNTHETIC LEATHER: A MECHANICAL VIEWPOINT FOR SUSTAINABILITY

JAMALUDDIN MAHMUD, SITI HAJAR MOHD YUSOP, and
SITI NOOR AZIZZATI MOHD NOOR

*Faculty of Mechanical Engineering, Universiti Teknologi MARA,
Shah Alam, Malaysia*

CONTENTS

OVERVIEW

Leather materials are widely used in many applications pertaining to fashion, which includes garments, handbags, shoes, belts, etc. Due to cost and availability, synthetic leather has become a popular material for replacement. This paper for the first time highlights previous studies and research work conducted on leather and synthetic leather. There is no denying that leather and synthetic leather development are very much important, especially in the fashion industry. This classification of material is produced in mass production to fulfill the consumers' requirement. This is because, apart from fashion related applications, leather and synthetic leather are also used for many other sectors such as automotive, textile, furniture, sports and others. More than 80% added value for leather can be attained, such as that seen in India, which has already achieved US$2 billion/year. The export of leather has increased constantly since the past decade and it has already been proven in research. Florsheim, Nunn Bush, Reebok, Stacy Adams, Gabor, Nike, Clarks, Slamander, Adidas, Ecco, Deichmann, Cole Hann, Elefanten, St Michaels and Wal Mart, are brands that make products from Indian leather. The amount billion dollars and brands confirm the importance of leather and synthetic leather materials. Initially, this paper focuses on previous studies that have been carried out to investigate the biomechanical properties. This was subsequently followed by the investigation of biomechanical properties on leather by integrating experiment-analytical-numerical approaches. Then finally, a novel framework is proposed to improve the biomechanical properties of leather and synthetic leather using hyperelastic material model for sustainability. The study is novel as no similar approach has been reported earlier.

9.1 PROPERTIES AND FUNCTIONS OF SKIN AND LEATHER

The ASTM D1517 states that skin is "the pelt of a small animal, such as calf, pig and sheep and also used interchangeably with hide." ASTM D1517 defines hide as "the pelt of a large animal, such as cow and horse, used interchangeably with skin." The standard also expressed leather as "a general term for hide or skin that still retains its original fibrous structure more or less intact, and that has been treated so as to be imputrescible even after treatment with water. The hair or wool may or may not have been removed. Certain skins, similarly treated or dressed, and without the hair removed, are termed "fur." No product may be described as leather if its manufacture involves breaking down the original skin structure into fibers, powder or other fragments by chemical or mechanical methods, or both, and reconstituting these fragments into sheets or other forms" [1]. The development of human skin assists to correlate study in animal skin as a references or datum for research of 'hide' to be employed for human usage. History shows that before 3000 B.C., the Chinese people started to use leather for their clothing [2]. For example, they used animal skin for coats, aprons and shawls. Therefore, it is important to develop the knowledge on leather for human purposes such as in current functions and also for the future. Compared to human skin, leather differs in terms of the layers it has. Animal skin only has two layers composed of different structures [3]. Grain layer is the upper layer and corium is the second inner layer. The definition of grain layers and corium are the "entire length of hair follicles and comprises mostly thick bundles of collagen fibers", respectively. In another perspective, leather is also defined as a collagen structure material. Recently, in manufacturing process, fat liquoring agents are usually used as additional material but this causes instability during heating session [4]. It should be avoided by improving the manufacturing process whilst maintaining leather quality. Natural leather has always been chosen to be made into leather goods products [5]. As evidence, it can be seen that there are numerous applications in downstream areas which apply leather as their intermediate industrial products. As mentioned before, India is one of the countries which exports leather in bulk quantity and this has already reached US$2 billion/year in the past decades [6].

Florsheim, Nunn Bush, Reebok, Stacy Adams, Gabor, Nike, Clarks, Salamander, Adidas, Ecco, Deichmann, Cole Hann, Elefanten, St Michaels and Wal Mart are some of global brands which source their materials from India. These brands that have contributed to billion dollars' worth to fashion, automotive, textile, furniture and sports industries confirm the importance of leather materials and thus proving the significance of this paper and the current review. Moreover, many researches claim that leather is economically significant because it is derived as a major by-product from the meat industry [7]. Thus, the current study is important as it could continuously support leather and synthetic leather research to provide a significant knowledge in understanding the biomechanical behavior of skin, leather and synthetic leather. This will help to further develop and improve leather and synthetic leather properties.

Before developing leather material, the process of the life cycle system of leather should be understood in-depth. Development of leather leads to more research to improve the strength and its quality. Leather has gained popularity because of its performance quality. The high price of original leather gives rise to an exclusive group of people who can afford to own the leather products in the form of belts, handbags, purses and others in their collection. The quality of leather depends on its composition. The more composition of leather in the product, the higher production costs will be. Based on past research, the improvement of leather is viewed from different perspectives which are in terms of manufacturing process. Past researches report that grafted leather samples by 2-ethyl hexyl acrylate (EHA) or butyl acrylate (BuA) are used to enhance the elasticity of leather fibers and also can act as a lubricant [8]. The leather grafting process helps to improve the leather tensile strength. Through experiments, improvements have been done using this lubricant to see the impact of its strength. The process starts from pesticide and fertilizer production. The process continues to two different sections that is the cultivation of crops and arable land. The cultivation process leads to feed production. From feed production and arable land, the process moves on to cattle rising. The slaughtering process begins and goes on to preservation. The next process is tanning and finishing which is followed by waste management and distribution. The leather is then distributed and used by humans [6]. The unwanted leather ended at the disposal section. This process needs energy, chemicals and water as the input to complete the process. Meanwhile the output of this process produces air emissions, waste water and solid waste. This process should be controlled to avoid pollution in the form of chemical usage to improve the leather quality. In contrast, tensile strength, percentage of elongation, tears strength and grain crack will not make any changes to leather strength when chromium in the tanning experiment was developed [9]. It has been proven that using chromium will not affect the leather quality.

9.2   LEATHER POTENTIAL

The potential of leather to be turned into belts or handbags very much depends on its quality, durability and esthetic values. The usage of leather is not only limited to accessories, it also is used in the footwear industry. The footwear industry has also been one of the top leather consumers. For this reason the hygiene issues have already created a new challenge for the suppliers to ensure leather quality as this is a consumer concern. In a hygiene study, a solution was achieved to solve the hygiene issues by using chitosan-based antimicrobial leather coatings development to eliminate *E. coli* from the animal skin [10]. Another interesting research on leather was made on gloves to handle bats. Cow, deer, elk, goat, and pig are the five types of leather glove used as the experiments sample. This study proved that deer skin characteristics have better dexterity and gloves made from deer leather are more supple than split-leather gloves [11]. As such, it is clear that different types of leather gave different performance depending on its application.

9.3   PROBLEM STATEMENT

Determining the accurate parameters of skin and leather has often been very difficult due to their complex structure. Skin behavior includes inhomogeneous, anisotropy and highly non-linear behaviors. This statement is supported by research which claimed that it is difficult to determine the identification of skin viscoelastic properties due to a large number of parameters [12]. Human skin is very complex and not easily modeled due to its mechanical behavior just like any other tissue [13]. Every layer in the skin has different mechanical properties. In recent years, development of skin characteristics have brought about some progress in the knowledge on skin but there are still no standard procedures that could be used as references for existing experimentation [14]. Viscoelastic, nonlinear and anisotropic behavior is some of the skin deformation areas which have been explored to obtain the experiment data for analysis by the researcher [15]. Additional materials were still chosen as upper materials to coat leather defects for many years [16]. It has been stated that the first synthetics were fabric coated with PVC, and the next material with PU. In physiology needs, synthetics hydrophilic properties, water vapor permeability and absorption still did not fulfill the need of the human foot requirement. The research also proved that PU synthetics are better at cold temperature, better handling and flexible compared to PVC synthetics. Leather quality has been the goal from early stage of manufacturing process. The question is how long will the quality of leather last. Figure 9.1 shows an example of belt defect after being used for a period of time.

FIGURE 9.1   Common belt defect (a) Front view (b) Rear view.

Each research of leather has their perspective and suggestions on how to improve the quality of material, from original leather until the process of synthetic leather. The researcher claimed that the most critical section in manufacturing process is in the final product such as bags, shoes and others [5]. In contrast, the development of sol-gel application for woods and textiles have been widely investigated compared to only a few cases which explored treatment of leather [17]. This statement supports the view that the development for better quality leather is still an on-going process.

9.4   SIGNIFICANCE OF STUDYING LEATHER

Studying skin and leather materials is one of the most critical challenges for research development. The behavior of skin is given priority in the mechanical view due to its wide application [18]. Furthermore, many engineering applications are in contact with human body [13]. The behavior of leather material has to be identified so that it can be controlled and improved for a better application.

In addition, medical imaging, face and gesture recognition are exam-ples of fields that are used to study the behavior of human body. It also increases the demand for non-rigid motion estimation especially in sci-ence area [19]. Leather becomes soft, malleable and suitability elastic. From the previous statement, past research demonstrate the presence of fat molecules for fat-liquoring process to isolate collagen fibers which help in improving the leather characteristic [20]. Besides wood and tex-tile, leather is also very important due to its high demand from many sectors such as in apparel, furniture or automotive [17]. For example, existing synthetics like Poromerics already have a niche in the footwear industry due to its special nature which can be engineered or tailored. The demand from users and manufacturers [16] make the research of quality of leather improve from time to time. In manufacturing issues, vacuum is one of the choices available to dry leather in a short time. The continuing of leather development is very important to allow for improvement of leather quality which needs detailed information for the drying process [21]. Recently, some manufacturers tend to use chromium-tanned leather which can cause health problems. The study of leather can be improved by monitoring the chromium content extracted from leather materials. The presence of chromium content is dangerous to humans especially in daily usage of leather textile materials [22]. This is why the leather properties should be understood from its basic structure and more study on what can replace the dangerous materials currently used on leather material. In consideration of human health, the presence of chromium should be replaced with other material that retains the same function in the manufacturing process of leather materials. From its original form as skin, then leather and later as synthetic leather, there is a need for more development to improve the properties of each stage. It is very important to study the material evolution so that quality of leather can be regenerated into a better and more durable product which can be used by consumers.

9.5   MECHANICAL TESTING AND STANDARDS

Testing of materials to determine its mechanical properties has been used in several studies. One of the research used an upgraded Instron mechanical property tester, model 1122 (Instron, Norwood, MA), and Test works 4 data acquisition software (MTS Systems Corp., Minneapolis, MN) to obtain the data for bovine wet white hides [7]. In addition, Universal testing Machine (Lloyd, LR 100 K) has been used to carry out modulus test according to ASTM D 790. The experiment was conducted at a crosshead speed of 50 mm/min to generate data for tensile strength, modulus and elongation according to ASTM D 638 [3]. Furthermore, a bespoke Zwick tensile testing machine (Zwick Testing Machines LTD, Herefordshire, UK), located in a laminar flow hood within a Containment 2 facility, was used to perform tensile test with a 5N load cell, with an accuracy of 0.4% [23]. Tensile tests were also performed by a Instron Mechanical test machine (Instron, Dynatup 9250) which used 240 N as load and conducted with 254 ± 50 mm/min of speed [24]. Therefore, tensile test is one of the common testing methods which have been used by past research to obtain the mechanical properties of materials.

Laboratory tests must be conducted according to proper standards to make sure the results are acceptable for the experiments. The ASTM International provides Standard Terminology Relating to Leather (D1517–10) [1], Standard Test Method for Elongation of Leather Designation (D2211–00) [25], Standard Test Method for Tensile Strength of Leather (D2099–00) [26], Standard Practice for Conditioning Leather and Leather Products for Testing (D1610–01) [27], and Standard Test Method for Measuring Thickness of Leather Test Specimens (D1813–13) [28]. As mentioned earlier, ASTM D 638 has been commonly used in the past researches to obtain tensile strength, modulus and elongation at a cross speed of 50 mm/min together with ASTM D790 for modulus test by using the Universal testing machine (Lloyd, LR 100 K) [3]. Apart from that, the ASTM D2209–00 has been applied to obtain stress-stretch data on bovine skin under uniaxial tension skin [24].

Mechanical and thermal properties of ABS resin have been carried out to reduce the cost, resource utilization and environment benefits. Due to the importance of the objective of this particular study, several ASTM International Standards have been referred to in conducting the study. For example, impact tester (ATS FAAR, Italy) was done according to ASTM D 256 A and B standards, respectively, in the standard laboratory atmosphere, Izod and Charpy impact strength tests were per-formed in Izod-Charpy Digital. Additionally, 100 mm diameter discs using CS–10 wheels in reference to ASTM D 1044 was used to measure abrasion resistance. 50 mm diameter discs according to ASTM D 785 are used to obtain Rockwell hardness. Extrudate material in reference to ASTM D 792 was performed to obtain Density. ASTM 648 and ASTM D 1525 in HDT–VICAT Testers (ATS FAAR Italy, model MP/3) were used to generate heat deflection temperature test and Vicat softening point tests, respectively. Dupont 910 series thermal analyzer system at the rate of 20°C/min from ambient to 800°C in nitrogen atmosphere was carried out to observe thermo-gravimetric analysis (TGA). An additional stan-dard used by previous research was ASTM D 570 standard test method showed increase of water absorption of 50 mm diameter disc specimens [3]. Furthermore, ASTM International, PN-EN ISO 3376:2005 was also used to perform tensile test. PN-EN ISO 3377–1:2005 was used to gain tear strength. Both of these standards were run using Zwick Rowell test-ing machine to investigate strength characteristics of the fat liquored leathers. PN-EN ISO 2418:2005 was used in leather samples for the physical testing was taken parallel to backbone from the leather samples [20]. In addition, ISO37 was used to perform tensile uniaxial stress ver-sus strain response for each silicone rubber at a strain rate of 0.3 s^{-1} measured according to the procedure detailed [29]. Many standards are available and experiment references were used to accomplish each aim of the research.

9.6 EXPERIMENTAL APPROACH

Specimen preparation began by determining the schematic diagram of the mechanical drawing dumbbell shape [26] with dimensions of 171 mm length and 31.8 mm width according to ASTM D2099–00 based on Figure 9.2. The goat leather specimens were divided into two directions which were longitudinal and transverse directions. Twenty specimens were prepared according to each direction. Forty specimens were prepared for this category.

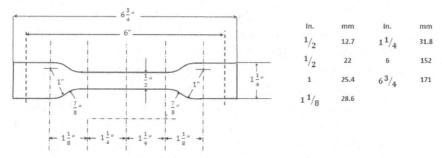

FIGURE 9.2   The specimen for tensile test according to ASTM D2099–00 [26].

A uniaxial tensile test was carried out using Universal Tensile Machine model INSTRON 3382 which is located in the Strength of Material Laboratory, Faculty of Mechanical Engineering UiTM as shown in Figure 9.3. The tensile testing was carried out according to ASTM D2099–00 standards. The goat leather was placed between the jigs and clamped using the jaws properly. Figure 9.4 displays the motion capture for uniaxial tensile test for goat leather. The image shows the motion from start until end of the test. The picture clearly illustrates that goat leather collapsed at the end of the test.

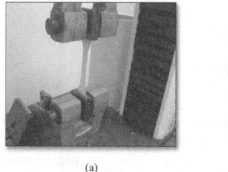 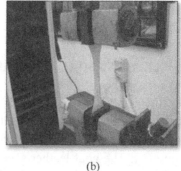

(a)   (b)

FIGURE 9.3   Different angles of specimen at uniaxial tensile test machine (a) Left view (b) Right view.

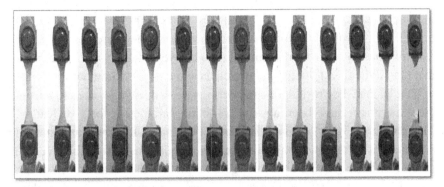

FIGURE 9.4 Motion capture of uniaxial tensile test for goat leather.

9.7 QUANTIFYING HYPERELASTIC PARAMETERS

The combination of experiment and hyperelastic material model was required to obtain biomechanical properties. With the strain value obtained in experimental procedures, the stretch value was calculated with a simple Eqs. (9.1) and (9.2):

$$\varepsilon = \lambda - 1 \tag{9.1}$$

$$\sigma_E = \frac{\mu}{\lambda}\left(\lambda^\alpha - \lambda^{-\frac{\alpha}{2}}\right) \tag{9.2}$$

where, ε = strain value; σ_E = engineering stress; λ = stretch value; μ, α = ogden parameter.

By having the stretch values [Eq. (9.1)], the simplified hyperelastic material model from the Ogden model then can be integrated to determine the Ogden parameter as shown in Eq. (9.2). Since the number of stretch values of the different stress level was calculated with one equation, then it could be computed simultaneously to get the specific Ogden parameter μ and α.

9.8 ANALYTICAL APPROACH

Experimentation by itself has not generated the biomechanical properties of goat leather. Regarding this issue, measuring goat leather by adapting

the analytical and numerical approaches to measure goat leather deformation is a great contribution to the success of the current study. This
research has applied Ogden models which were determined by experiment, analytical and numerical approaches to analyze parameter presence
in goat leather. The results of engineering stress, σ_E against stretch, λ from
the experiment have been applied to obtain parameters for goat leather.
The parameter was obtained for analytical approach by mathematical
equation using algebra and simultaneous equations. After the average
parameter has been gathered for Ogden (μ and α), the parameter has been
reused to gather engineering stress, σ_E against stretch, λ for goat leather.
Next, the values of engineering stress, σ_E (MPa) against stretch, λ data
was used for graph plotting for both data which includes experimental and
analytical approach.

9.9   NUMERICAL APPROACH

Numerical approach has been applied by curve fitting technique to determine the parameters for goat leather. It began by inserting data of engineering stress, σ_E and stretch, λ into hyperelastic material models for the
Ogden (μ and α) model. Next, the values of engineering stress, σ_E against
stretch, λ data were used for graph plotting for both data including the
experiment and numerical approach. The curve fitting technique used
a program from Microsoft Excel. By using this Microsoft Excel, a program to find parameters in Ogden (μ and α) including longitudinal and
transverse directions was conducted. The result of engineering stress, σ_E
against stretch, λ from goat leather was used to determine the parameter
for Ogden (μ and α) model. The engineering stress, σ_E and stretch, λ data
were gathered from the experiment. The step continued with the difference
in engineering stress, σ_E and predicted stress, σ_E value. The differences
were calculated for the square of differential value. Next, the square of the
differences was summed up (total) using AutoSum Formula. After all the
data was gathered, the solver parameters were chosen to solve the parameter determination. The target cell was filled with the sum of differences
and the changing cells had been inserted by parameters. The solve button
was used and the parameter was defined. The difference between predicted
stress, σ_E and engineering stress, σ_E had been minimized by changing the
value of μ and α. Both of the engineering stress, σ_E and predicted stress, σ_E

values were plotted according to the parameter obtained from the solver parameter. The predicted parameter value, sometimes did not fall on the engineering stress curve, which means the predicted parameters were not good. To solve this issue, the parameter had to be closer to the final predicted value which is a repeated step by using solver until the best result can be determined.

9.10 RESULTS AND DISCUSSION

The results of engineering stress, σ_E against stretch, λ from the experiment were applied to obtain parameters for goat leather. The maximum load applied to the experiment was, according to ASTM D2099–00 is 222 N. However, for goat leather, the maximum load was only plotted until 140 N due to the specimens' average breaking point (specimen split into two sections) during the experiment. At 140 N of load, the maximum extensions were plotted until 60 mm for forty specimens of the total specimens. This result for goat leather includes longitudinal and transverse directions. Table 9.1 shows the data which was tabulated to facilitate the comparison between analytical and numerical data. The y-axis represents engineering stress, σ_E (MPa) data and x-axis represents stretch, λ data extracted from experiment, analytical and numerical approaches shown in Table 9.1 are the curve plots, according to the experimental data, analytical data and numerical data. For goat leather biomechanical properties which have been determined using an analytical approach for Ogden parameter, (α and μ) are from 12.942 to 19.779 and from 1.178 MPa to 1.789 MPa, respectively. Based on Table 9.1, it shows that Ogden parameter (μ and α) using an analytical approach does not fit the experimental curve as the analytical curve is significantly different from the experimental curve. For goat leather biomechanical properties, which was determined using a numerical approach for Ogden parameter, (α and μ) are 15.885 and from 1.827 MPa to 4.441 MPa, respectively. Between both directions, the longitudinal direction is stretchier than transverse direction. Table 9.1 shows that the Ogden parameter for numerical approach fits closely with the experimental data. As a result, the material parameters for numerical approach closely fit to represent the experimental data as compared to the analytical approach for both directions.

TABLE 9.1   Parameters of Goat Leather for Ogden (μ and α) for Longitudinal and Transverse Directions

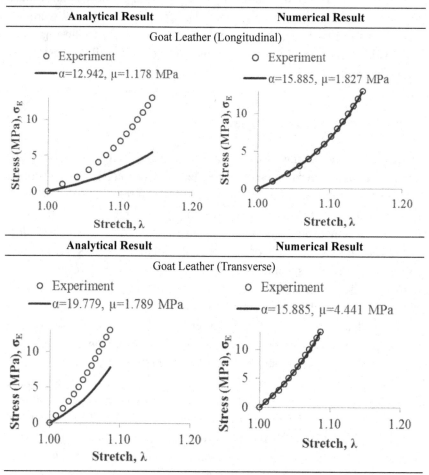

9.11   COMPARISON WITH OTHER MATERIALS

Table 9.2 clarifies the results of biomechanical properties determined from the analytical and numerical approaches to the data obtained from a tensile test for goat leather. Based on Table 9.2, μ value is higher for the numerical approach compared to the analytical approach. Meanwhile, for α value, the numerical approach parameter is in between the analytical approach parameter. As shown in Table 9.2, the biomechanical properties of goat leather was in between circular thermoplastic membranes and single wall

TABLE 9.2 The Ogden Parameters, α Values and μ

Authors	Order of Magnitude		Method	Specimen
	Shear Modulus, μ	Strain Hardening Exponent, α		
Mahmud et al. [30]	26	110	Finite Element Models	Human Skin
Mahmud et al. [14]	10	110	Parametric Study and Simulation	Human Skin
Manan et al. [15]	10	40	Finite Element Modeling and Simulation	Human Skin
Shergold et al. [29]	0.4	12	Uniaxial Stress	Pig Skin (strain rate 0.004 s⁻¹)
Manan et al. [31]	0.4	4.6	Uniaxial Tension	Bovine Skin
Chang et al. [32]	0.028	4.391	Uniaxial Tension	Rubber Material
	5.813	0.031		
	0.953	2.593		
Brink and Stein [33]	0.63	1.3	Finite Element Computations	Rubber Sealing
	0.0012	5		
Elango et al. [34]	0.029	3.123	Uniaxial Tensile Test	Silastic P–1 Silicone
Beatty [35]	1.491	1.3	Membrane Inflation Pressure	Rubber
	0.003	5		
	−0.0237	−2		

TABLE 9.2   Continued

Authors	Order of Magnitude		Method	Specimen
Rashid et al. [36]	0.09 (1/s)	2.5	Compression Test	Fresh Porcine Brain
	0.18 (30/s)	2.7		
	0.18 (60/s)	3.3		
	0.20 (90/s)	3.0		
Lim et al. [37]	10/3 (0.005/s)	11/7	Uniaxial Tension	Pig Skin
	20/8 (0.5/s)	11/7		
	180/40 (1700/s)	11/7		
	230/200 (2500/s)	11/7		
	300/370 (3500/s)	11/7		
Zhiqiang and Scanlon [38]	11.1 kPa	5.6	Compression Test	Bread crumb of the hyperfoam
	−8.1 kPa	2.8		
Rashid et al. [39]	3.076 kPa (37°C)	2.94	Compression Test	Fresh porcine brain
	4.824 kPa (37°C)	3		
	3.205 kPa (22°C)	3.08		
	4.575 kPa (22°C)	4.53		
Ayoub et al. [40]	0.03 MPa	11×10^3	Fatigue Test	Styrene-butadiene rubber filled with 34 parts per hundred rubber
	4.03 MPa	10×10^{-4}		

TABLE 9.2 Continued

Authors	Order of Magnitude		Method	Specimen
Kakavas [41]	−2.013 MPa	2.792	Uniaxial Tension	Ethylene Propelene Diene Terpolymer (EPDM) elastomer filled with carbon black
	2.113 MPa	2.364		
	0.0619 MPa	3.223		
Erchiqui and Kandil [42]	0.0631 MPa (150°C)	1.968	Inflation Test	Circular Thermoplastic Membranes
	0.4820 MPa (145°C)	1.601		
	−2.5687 MPa (150°C)	−0.5726		
	0.117 MPa (145°C)	1.004		
	0.2085 MPa (150°C)	1.5757		
	0.4115 MPa (150°C)	0.0243		
	0.1827 MPa (150°C)	0.0481		
	5.9528 MPa (145°C)	0.5894		
	0.1926 MPa (145°C)	2.0689		
	−0.1288 MPa (145°C)	−0.8396		
Current (Numerical)	**1.827 MPa to 4.441 MPa**	**15.885**	**Tensile**	**Goat Leather**
Current (Analytical)	**1.178 MPa to 1.789 MPa**	**12.942 to 19.779**	**Tensile**	**Goat Leather**
Flores et al. [43]	217.99 GPa (0.34 nm)	−10	Finite Element Based Lattice Approach	Single Wall Nanotubes
	435.989 GPa (0.34 nm)	−14		
	5785.124 GPa (0.066 nm)	−10		
	11570.25 GPa (0.066 nm)	−14		

nanotubes parameters. Both of the parameters obtained from the analytical and numerical approaches was greater than other materials such as human skin, pig skin and fresh porcine brain.

9.12   PROPOSED FRAMEWORK: LEATHER FOR SUSTAINABILITY

This paper proposed a novel framework for investigating the biomechanical properties of goat leather by integrating experiment-analytical-numerical approaches for sustainability. This is unique because no similar approach has been reported before in conducting research in goat leather biomechanics. The quality of leather or synthetic leather development has improved from the manufacturing process and human purpose. None of the previous researches contributed to determining the parameters of goat leather material. Based on the literature and the research gap discovered in the previous sections, the proposed framework started from the composite phase as in Figure 9.5. The next stage will continue with the bio-composite phase, the combination of biomaterial (e.g., Leather) and other materials (e.g., PU synthetic). The bio-composite phase will go through a mechanical testing, which will provide the curve for the bio-composite material. The curve for bio-composite material will represent the biomechanical properties as a benchmark to improve the properties of bio-composite material. Each of the samples will be tested in the laboratory -using tensile testing machine, Instron Mechanical test machine (Instron, Dynatup 9250). The data gained will be useful for the mechanical behavior study of leather using stretch-stress analysis. Finally, the best result for biomechanical properties will produce leather which is sustainable for a longer period of time than the current available synthetic leather. Figure 9.6 shows the hypotheses for the curve improvement (from curve A to curve B) for leather and synthetic leather properties. This improvement will make the leather and synthetic leather softer (less stiff) and can resist higher stress, which can improve the sustainability of the leather and synthetic leather for product usage.

FIGURE 9.5   Proposed framework for sustainability.

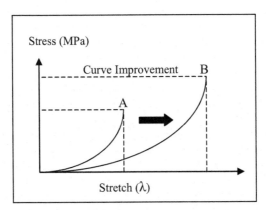

FIGURE 9.6   Curve improvement (from Curve A to B) for leather and synthetic leather properties.

9.13   CONCLUSIONS

This paper discussed the functions, properties, problems, issues, challenges, and the potential application of skin, leather and synthetic leather. Previous researches have been reviewed to highlight the importance of pursuing the current study. This paper proposed further investigation of the bio-mechanical properties of leather by integrating experiment-analytical-numerical approaches. Finally, a novel framework is proposed for this study to improve the mechanical properties of synthetic leather using hyperelastic material model for sustainability. The study is distinctive as no similar approach has been reported earlier. The results and findings of the study using the current approach will be reported in the near future. The gained knowledge will be significant in improving the mechanical behavior of leather and synthetic leather.

ACKNOWLEDGMENT

This research was sponsored by the Ministry of Education (MOE) of Malaysia and Universiti Teknologi MARA (UiTM) Malaysia, grant no FRGS/1/2012/TK01/UITM/02/4 (UiTM File. No. 600-RMI/FRGS 5/3 (25/2012)) and Grant No: 600-RMI/ERGS 5/3 (16/2012).

KEYWORDS

- biomechanical
- biomechanics
- leather
- skin
- sustainability
- synthetic leather

REFERENCES

1. *Standard Terminology Relating to Leather, in Designation: D1517 – 10*, ed. 100 Barr Harbor Drive, PO Box C700, West Conshohocken, PA 19428–2959. United States: ASTM International. (2013).
2. Luo, W., Si, Y., Wang, H., Qin, Y., Huang, F., & Wang, C. (2011). Leather Material Found on a 6th B. C. Chinese Bronze Sword: A Technical Study. *Spectrochimica Acta Part A: Molecular and Biomolecular Spectroscopy. 79*, 1630–1633.
3. Ramaraj, B. (2006). Mechanical and Thermal Properties of ABS and Leather Waste Composites. *Journal of Applied Polymer Science. 101*, 3062–3066.
4. Bajza, Z., & Vrcek, I. V. (2001). Fat liquoring Agent and Drying Temperature Effects on Leather Properties. *Journal of Materials Science. 36*.
5. Failli, F., & Dini, G. (2004). An Innovative Approach to the Automated Stacking and Grasping of Leather Plies. *CIRP Annals – Manufacturing Technology. 53*, 31–34.
6. Joseph, K., & Nithya, N. (2009). Material Flows in the Life Cycle of Leather. *Journal of Cleaner Production. 17*, 676–682.
7. Liu, C.-K., Latona, N. P., Ramos, M. A., & Goldberg, N. M. (2010). Mechanical Properties and Area Retention of Leather Dried with Biaxial Stretching Under Vacuum. *Journal of Materials Science. 45*, 1889–1896.
8. El-Ghaffar, M. A. A., El-Sayed, N. H., & Masoud, R. A. (2003). Modification of Leather Properties by Grafting. I. Effect of Monomer Chain on the Physico-Mechanical Properties of Grafted Leather. *Journal of Applied Polymer Science. 89*, 1478–1483.
9. Kanagaraj, J., Chandra Babu, N. K., & Mandal, A. B. (2008). Recovery and Reuse of Chromium from Chrome Tanning Waste Water Aiming Towards Zero Discharge of Pollution. *Journal of Cleaner Production. 16*, 1807–1813.
10. Fernandes, I. P., Amaral, J. S., Pinto, V., Ferreira, M. J., & Barreiro, M. F. (2013). Development of Chitosan-Based Antimicrobial Leather Coatings. *Carbohydr Polym.* Oct 15 *98*, 1229–35.
11. Freeman, P. W., & Lemen, C. A. (2009). Puncture-Resistance of Gloves for Handling Bats. *Journal of Wildlife Management. 73*, 1251–1254.

12. Delalleau, A., Josse, G., Lagarde, J. M., Zahouani, H., & Bergheau, J. M. (2008). A Nonlinear Elastic Behavior to Identify the Mechanical Parameters of Human Skin in Vivo. *Skin Res Technol.* May *14*, 152–64.

13. Holt, C. A., & Evans, S. L. (2009). Measuring the Mechanical Properties of Human Skin in Vivo Using Digital Image Correlation and Finite Element Modeling. *The Journal of Strain Analysis for Engineering Design. 44*, 337–345.

14. Mahmud, J., Holt, C., Evans, S., Manan, N. F. A., & Chizari, M. (2012). A Parametric Study and Simulations in Quantifying Human Skin Hyperelastic Parameters. *Procedia Engineering. 41*, 1580–1586.

15. Manan, N. F. A; Ramli, M. H. M; Patar, M. N. A. A; Holt, C., Evans, S., Chizari, M., et al. (2012). Determining Hyperelastic Parameters of Human Skin Using 2D Finite Element Modeling and Simulation. *IEEE Symposium on Humanities, Science and Engineering Research.*

16. Merkle, R., & Tackenberg, W. Alternatives to Leather: Laif Poromerics A New Family of Polyurethane Coated Fabrics. *Journal of Industrial Textiles.* (1984). *13*, 228–238.

17. Mahltig, B., Vossebein, L., Ehrmann, A., Cheval, N., & Fahmi, A. (2012). Modified Silica Sol Coatings for Surface Enhancement of Leather. *Acta Chim. Slov.* 331–337.

18. Mahmud, L., Ismail, M., Manan, N. (2013). Characterization of Soft Tissues Biomechanical Properties Using 3D Numerical Approach. *(BEIAC), 2013 IEEE.*

19. Tsap, L. V., Goldgof, D. B., & Sarkar, S. (1997). Human Skin and Hand Motion Analysis from Range Image Sequences Using Nonlinear FEM. *IEEE.*

20. Żarłok, J., Śmiechowski, K., Mucha, K., & Tęcza, A. (2013). "Research on Application of Flax and Soya Oil for Leather Fat Liquoring." *Journal of Cleaner Production.*

21. Barni, R., Zanini, S., Piselli, M., & Riccardi, C. (2006). Experimental Study of the Behavior of Leather Under Vacuum Conditions. *Vacuum. 81*, 265–271.

22. Rezić, I., & Zeiner, M. (2008). Determination of Extractable Chromium from Leather. *Monatshefte für Chemie – Chemical Monthly. 140*, 325–328.

23. Groves, R. B., Coulman, S. A., Birchall, J. C., & Evans, S. L. (2013). An Anisotropic, Hyperelastic Model for Skin: Experimental Measurements, Finite Element Modeling and Identification of Parameters for Human and Murine Skin. *Journal of the Mechanical Behavior of Biomedical Materials. 18*, 167–180.

24. Manan, N., Mahmud, J., & Ismail, M. (2013). *Quantifying the Biomechanical Properties of Bovine Skin under Uniaxial Tension: jomb.org.*

25. *Standard Test Method for Elongation of Leather, in Designation: D2211-00,* ed. 100 Barr Harbor Drive, PO Box C700, West Conshohocken, PA 19428–2959. United States: ASTM International. (2010).

26. *Standard Test Method for Tensile Strength of Leather, in Designation: D2209-00,* ed. 100 Barr Harbor Drive, PO Box C700, West Conshohocken, PA 19428–2959. United States: ASTM International. (2010).

27. *Standard Practice for Conditioning Leather and Leather Products for Testing, in Designation: D1610 – 01,* ed. 100 Barr Harbor Drive, PO Box C700, West Conshohocken, PA 19428–2959. United States: ASTM International. (2013).

28. *Standard Test Method for Measuring Thickness of Leather Test Specimens, in Designation: D1813 – 13,* ed. 100 Barr Harbor Drive, PO Box C700, West Conshohocken, PA 19428–2959. United States: ASTM International. (2013).

29. Shergold, O. A., Fleck, N. A., & Radford, D. (2006). The Uniaxial Stress Versus Strain Response of Pig Skin and Silicone Rubber at Low and High Strain Rates. *International Journal of Impact Engineering. 32*, 1384–1402.

30. Mahmud, L., Manan, N. F. A., Ismail, M. H., & Mahmud, J. (2013). Characterization of Soft Tissues Biomechanical Properties Using 3D Numerical Approach. *IEEE Business Engineering and Industrial Application Colloquium (BEIAC)*.

31. Manan, N. F. A., Mahmud, J., & Ismail, M. H. (2013). Quantifying the Biomechanical Properties of Bovine Skin under Uniaxial Tension. *Journal of Medical and Bioengineering. 2*, 45–48.

32. Woo, C. S., Park, H. S., & Shin, W. G. (2011). Finite Element Analysis by Using Hyper-Elastic Properties for Rubber Component. *Key Engineering Materials. 488–489*, 190–193.

33. Brink, U., & Stein, E. (1998). A Posteriori Error Estimation in Large Strain Elasticity using Equilibrated Local Neumann Problems. *Computer Methods in Applied Mechanics and Engineering. 161*, 77–101.

34. Elango, N., Mohd Faudzi, A. A., Muhammad Razif, M. R., & Mohd Nordin, I. N. A. (2013). Determination of Non-Linear Material Constants of RTV Silicone Applied to a Soft Actuator for Robotic Applications. *Key Engineering Materials. 594–595*, 1099–1104.

35. Beatty, M. F. (2011). Small Amplitude Radial Oscillations of an Incompressible, Isotropic Elastic Spherical Shell. *Mathematics and Mechanics of Solids. 16*, 492–512.

36. Rashid, B., Destrade, M., & Gilchrist, M. D. (2012). Determination of Friction Coefficient in Unconfined Compression of Brain Tissue. *Journal of the Mechanical Behavior of Biomedical Materials II. Oct 14*, 163–171.

37. Lim, J., Hong, J., Chen, W. W., & Weerasooriya, T. (2011). Mechanical Response of Pig Skin Under Dynamic Tensile Loading. *International Journal of Impact Engineering. 38*, 130–135.

38. Liu, Z., & Scanlon, M. G. (2003). Modeling Indentation of Bread Crumb by Finite Element Analysis. *Biosystems Engineering. 85*, 477–484.

39. Rashid, B., Destrade, M., & Gilchrist, M. D. (2012). Temperature Effects on Brain Tissue in Compression. *Journal of the Mechanical Behaviour of Biomedical Materials. Oct 14*, 113–118.

40. Ayoub, G., Naït-abdelaziz, M., Zaïri, F., & Gloaguen, J. M. (2010). Multiaxial Fatigue Life Prediction of Rubber-Like Materials Using the Continuum Damage Mechanics Approach. *Procedia Engineering. 2*, 985–993.

41. Kakavas, P. A. (2001). Evaluation of the Derivatives of the Strain Energy Function with Respect to Strain Invariants for Carbon Black Filled EPDM. *Polymer Engineering and Science. September 41*, 1589–1596.

42. Erchiqui, F. (2006). Neuronal Networks Approach for Characterization of Softened Polymers. *Journal of Reinforced Plastics and Composites. 25*, 463–473.

43. Saavedra Flores, E. I., Adhikari, S., Friswell, M. I., & Scarpa, F. (2011). Hyperelastic Finite Element Model for Single Wall Carbon Nanotubes in Tension. *Computational Materials Science. 50*, 1083–1087.

CHAPTER 10

WATER BASED PACK CARBURIZING TECHNIQUE: A HEAT TREATMENT PRACTICE FOR SUSTAINABLE ENVIRONMENT

MUHAMMAD HAFIZUDDIN JUMADIN,[1] BULAN ABDULLAH,[1] MUHAMMAD HUSSAIN ISMAIL, and SITI KHADIJAH ALIAS[2]

[1]*Faculty of Mechanical Engineering, Universiti Teknologi MARA, Shah Alam, Malaysia*

[2]*Faculty of Mechanical Engineering, Universiti Teknologi MARA, Johor, Malaysia*

CONTENTS

OVERVIEW

Steel making process is familiar with environmental issues (high energy usage, carbon emission, pollution, etc.) but yet, the production demands keep rising as for the development of the industry nowadays [1, 2]. The metallurgists come up with heat treatment techniques that apply to steel in order to control the issues, a technique which increases the durability and prolongs the product life of the steel since the raw material of the steels is a non-renewable resource and keep depleting. Heat treatment by pack carburizing is the alternative to keep the product life cycle sustained. The pack carburizing is the most environmental friendly compared to the other steel heat treatment (low energy usage, less contaminate waste) but the carbon diffusion, which represent the efficiency of the hardness properties of the product were often limited [3]. Thus the research by using alteration of the pack carburizing medium to increase the carbon dispersion was found by using paste, a water based pack carburizing medium. By using paste medium, it can increase the life of the steels and contributed toward the sustainable environment.

10.1   INTRODUCTION

Parts of sustaining the environment are by preserving the nature of the resource. The action can be done by using the source properly and increase the product's potential so it can last longer. So, the materials for producing the end product will not be wasted. Carburizing is one of the process for steel product (such as gear and shaft) that occasional execute the idea of expanding the lifetime of the product. Steel is used widely by human in their daily life. Steels were used widely, from a small purposed such as hairpin to a mega development, such as building's beam and infrastructure. But the carburizing process is more specifically suitable for a steel product that requires high wear resistance and hardness since this type of application were regularly failed and required maintenance. When the product failed, it needs to be replaced with new one, and it will be such a waste from all kinds of aspects. So the best solution for this problem is by applying the carburizing treatment toward the product's properties enhancement.

10.2 CARBURIZING

Carburizing treatment has been discovered for so many years and it had developed from time to time in order to increase its efficiency of making the steel product long lasting [5]. Although, it has come to advance method, the basic idea stays the same, to increase the carbon content (carbon provides hardness properties to steel) on the surface of the steel so it will become hard and preserved the ductility of the steel core. This method requires three simple rules [5, 6]:

1. The steel supplied with the high carbon rich environment.
2. The austenitic temperature is required in order for the steel to allow the diffusion process.
3. A soaking time is required for the carbon element to disperse inside the steel surface toward the core.

There are several types of carburizing, but the common used were powder (pack) carburizing, fluid carburizing and gas carburizing [7]. The differences between these carburizing are their carbon compound used (powder, liquid and gases). The cheapest and easiest type is the pack carburizing. It is environmentally friendly compared to fluid carburizing and it is way cheaper and easily operates compared to gas carburizing. The problem with pack carburizing, it requires high temperature and high soaking time. High temperature and high soaking time mean high energy needed for combustion of the furnace for a longer time and this will contribute to destroy of environment and waste of natural resources.

10.3 WATER BASED PACK CARBURIZING

Paste, a mixed substance between solid and liquid, which offered flowability, thicker and more affordable process of carburizing process is one of newly discover method which it might bring an intersession property of solids and liquid. Previous researcher [6] have applied a paste of carbon and boron compound for the heat treatment. The result shows paste improves 27% better case depth compared to conventional pack hardening using powder. Paste acts as self-protective film, which theoretically will prevent oxidation to the part. The diffusion of carbon into the surface affects the density of the steel. To ensure proper carbon dispersion, the temperature was maintained at high range. The steel is heated to a temperature at least 800°C and above, the composition of the molecules and the volume is restructured with the diffusion of carbon content which affected the hardness and density at once [7]. Several experiments and testing were conducted to validate the efficiency of the water based pack carburizing gas compared to conventional pack carburizing. This article was explored on the microstructure and the hardness of water based pack carburized steel towards the dispersion of carbon from the surface toward the core.

10.4 WATER BASED PACK CARBURIZING PROCEDURES

Low carbon steel was used for this study in order to focus on increment of the carbon content in the steel. ASTM 850 Grade 70 steels were used as samples for the whole experiment. Samples were prepared with dimension of 2 mm width x 2 mm length x 2 mm height. Figure 10.1 shows the process flow of making the water based pack carburizing.

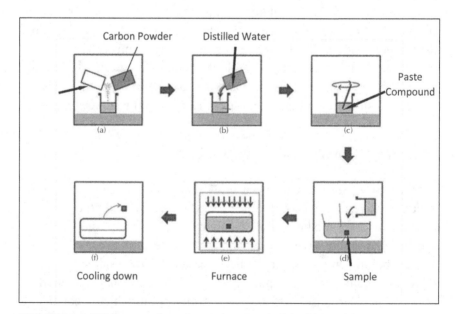

FIGURE 10.1   The process flow of water based pack carburizing.

TABLE 10.1   Ratio of Water to Powder for Paste Sample

Sample	Water	Powder
1:1	1	1
3:1	3	1
5:1	5	1
7:1	7	1
9:1	9	1

There are 6 parameters of paste used in the experiments. The ratio of water to the powder were shown in Table 10.1. In order to make the water based compound, activated charcoal powder was mixed with Sodium Carbonate (Na_2CO_3) and Barium Carbonate ($BaCO_3$). Then, mixed powder was diluted with distilled water according to specify weight ratio and became paste compound. The paste compound was stirred with the mixer until completely uniform. Samples were placed in a steel box. Next step, paste compound poured into the box until covers the samples. The steel case was sealed to keep the carbon element inside. The steel box heated up to 900°C for 8 hours. The sample was allowed to cool at room temperature and prepared for testing.

10.5   EXPERIMENTAL PROCEDURE

The samples were hot mounted and grind with sand paper grit from 120–1000. Samples were polished with diamond paste (size 9, 5, 3 and 1 micron). The samples were etched with Nital 5% before observed under optical microscopy, Olympus BX41M. The samples were indented with microhardness Vickers tester with load of 1 kg from the surface through the interior. The samples were indented at least 3 times with different places was taken in order to get average data for the hardness.

10.6   RESULTS AND DISCUSSION

Carbon layer was produced during the carburizing process. This case depth carries hardness properties which contributed to the mechanical properties of the steel. Case depth influence to wear resistance and hardness properties. When the wear resistance was high, it can conserve the life cycle of the steels. In order to observe the carbon diffusion, the sample needs to be observed under optical microscope. Microhardness test was carried out in order to validate the hardness properties which carried by the carbon element of the carburizing process. Higher hardness reading proved that higher carbon content toward the carburized steel surface.

10.6.1 OPTICAL MICROSTRUCTURE

The comparison between cross sections of sample's microstructure captured under optical microscope with difference weight ratio of water to powder as shown in Figure 10.2.

The untreated sample shows there was no layer formed except the microstructure of ferrite. The layer of carbon from carburized steel using fine powder was thicker compared to others samples. The layer was measured and the length was approximately 0.04 mm. Sample 1:1 was carburized by using paste compound, a mixture weight ratio of 1 carburizing powder compound with 1 distilled water.

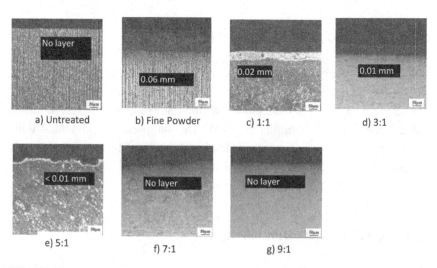

FIGURE 10.2 Microstructure of case depth layer: (a) untreated, (b) fine powder, and at different weight ratio of water to powder, (c) 1:1, (d) 3:1, (e) 5:1, (f) 7:1, (g) 9:1.

The results exhibited that water based carburized samples exhibited lower layer thickness compared to powder carburized sample where sample 3:1 indicated 0.01 mm and sample 1:1 is 0.02 mm. As the ratio of water increased, the thick layer decreased as shown in sample 5:1 where less than 0.01 mm was formed, and for sample 7:1 and 9:1 the thick layer became less visible. The information leads to clarification, as the distilled water content increased, the layer thickness of the carburized steels decreased. The same factors were discussed by Chen [9], on the effect of water (H_2O) content to the carburizing process.

The water content increased for the compound, the carbon layer formation decreased at the surface of carburized steels as shown in Figure 10.3. At one point when there was too much water diluted with the powder, the formation of a layer does not form as the carbon concentration was very low. The high carbon concentration leads to the rising of CO production during the process. Increasing carbon represent escalating the factor on the effectiveness for carburizing process.

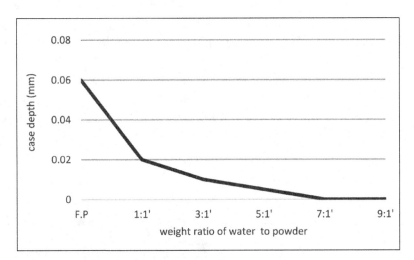

FIGURE 10.3   Case depth at different weight ratio of water to powder.

10.6.2   MICROHARDNESS

Carbon was known as it carries out the hardness properties toward the composition of the material [10]. The diffusion of the carbon in the carburized steel can be measured by inspection of the microhardness of the surface through the interior of the steel. Figure 10.4 shows the diamond shape Vickers across the sample indenter on the surface of the samples in order to recognize the diffusion of the carbon on the carburized steel. Case depth was not formed above the surface, but on the substrate that were formed on the surface of the steel. The result proved the size of the diamond shaped that formed in the area was essentially larger and hardness value was small, which is 100 HV. Carbon diffusion was measured under the deposit layer. The diamond-shape that formed a smaller size indicates higher hardness value [10]. The size of the indention was small for a certain amount of indentation. The size increased as it achieved at the depth where the carbon diffusion were limited. The verification of case depth of carbon diffusion was made. The same layer is shown in studies by Lou [6], on identification of effective carburized case depth.

The samples were measured the hardness in order to analyze the influences of depth of carbon rich diffusion on different type of compound as shown in Figure 10.5. The hardness value of the untreated sample between 118 and 140 HV was plotted as a benchmark to distinguish the increasing of the hardness after the carburizing process. Most of the samples showed a noticeable increment of hardness value at start, around 260–270 HV except for sample 9:1 which only had the value of 155 HV. The result showed that at 0.4 mm depth from the surface, all the hardness value of the samples were approximately at the 150–120 HV which similar to the hardness values of untreated sample.

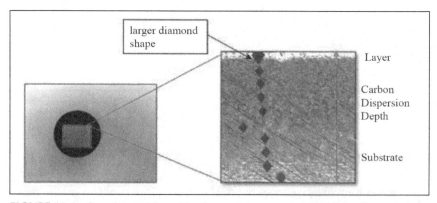

FIGURE 10.4 Samples position of hardness measurement and Vickers diamond shape on carburized steels.

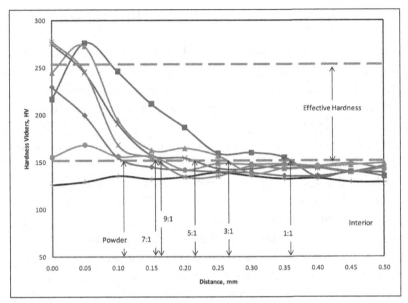

FIGURE 10.5   Hardness of carburized steel at different surface depth.

It can be specified that the carbon were diffused when the value of the hardness are above 150 HV. The powder carburizing sample shows that at 0.10 mm, the hardness value dropped to 150 HV. From the Figure 10.5, it shows that ratio 1:1 and ratio 3:1 have the deepest case depth which at 0.25 mm and 0.35 mm, hardness value was above 200 HV. For samples 7:1 and 5:1, it was shown that the hardness decreased at about 0.10 mm from the surface. For sample 9:1, hardness value was constantly between 150 and 145 HV along the depth of the substrate. The same trend showed in the study by Lou [6], where paste provides more deep diffusion compare to powder compound. A paste compound formed a self-protective layer which provides a protection for carbon gases to trap between the surface and compound. The gases will have much more chance to induce inside the steel rather than being released in other places.

10.7   CONCLUSION

The procedures for water based pack carburizing (or paste carburizing) are quite similar to conventional pack carburizing, except the water based pack carburizing used paste compound (which were the mixture of charcoal powder, $BaCO_3$ and Na_2CO_3 with the distilled water). It is easy to operate without requiring high skill workers. The compound is safe and can be managed easily comparing with bed fluid carburizing (which were considered as unfriendly to the environment because of the waste management were hazardous and not easily disposed) and gas carburizing (because of the expensive cost). So it verified that the technique promoted safe and sustainable environment.

From the experiment, it was found that the different carburizing compounds influenced the dispersion layer formation and the hardness through the carburized steels. By only using powder compound alone, it will barely increase the hardness at a certain case depth. Adding liquid to the compound and converts it into a paste (optimum weight ratio is 1:1 to 1:3 or less), will generate the deeper diffusion of carbon into the steel. Water based reduced the activation energy for carbon powder [15] and created more CO for carburizing adsorption [8]. Adding a large amount of liquid will reduce the concentration of carbon compound, so, an optimum content of liquid were essential in order for paste compound to work well. The production of carbon dioxide will be less because the concentration of carbon was distressed by dilution of higher water content [9]. Temperature and time require were less for water based pack carburizing comparing with conventional pack carburizing [6]. Since the conventional pack carburizing need additional time and temperature for the carbon diffusion to achieve the same level as water based pack carburizing in term of dispersion length.

The darker layer in between the upper surface and core surface is the one that carries hardness properties of the carbon. It can be observed that the layer uniform has the same length of the hardness versus depth profile. It proved that a higher concentration of the carbon paste compound will provide a higher thickness of the layer compare to the less carbon concentration paste compound. The experiment proved that an optimum weight ratio for water based pack carburizing were crucial in order to increase the efficiency of carbon diffusion on the surface of the steel.

From this research, the results showed that the water based pack-carburizing fit the purpose of sustainability of environment and natural resources. It can be concluded that:

1. Carburizing contributed to the enhancing the life cycle of low carbon steel, natural resources such iron are not wasted and less production cost of steelmaking.

2. Pack Carburizing are environmentally friendly compared with fluid carburizing (issue of waste disposal) and gas carburizing (expensive).
3. Water Based Pack Carburizing is an alternative way of pack carburizing which promoted high dispersion of carbon that requires less energy and time.

Water based pack carburizing is the alternative method of providing long lasting life cycle for steels. The environment can be preserved and the natural resources can be used wisely if this innovative heat treatment applies to steel.

ACKNOWLEDGMENT

This research was funded by the Ministry of Education (MOE), Malaysia and Universiti Teknologi MARA, Malaysia, mainly via grant no. GRANT 600-RMI/DANA 5/3/RIF (320/2012) and partly supported via grant no. 600-RMI/RAGS 5/3 (50/2013).

KEYWORDS

- case depth
- environmentally friendly
- low carbon steel microhardness
- pack carburizing
- water based

REFERENCES

1. Rajan, T. V., Sharma, C. P., & Sharma, A. (2004). *Heat Treatment Principle and Technique*, Prentice Hall of India.
2. Elzanaty, H. (2014). Effect of Carburization on the Mechanical Properties of the Mild Steel, *Innovative Space of Scientific Research Journals*, 6, 987–994.
3. Aramide, F. O., Ibitoye, S. A., Oladele, I. O., & Olatunde, J. B. (2009). Effects of Carburization Time and Temperature on Mechanical Properties of Carburized Mild Steel, Using Activated Carbon as Carburizer. *Material Research. 12*, 483–487.

4. Dragomir, D., Cojocaru, M. O., & Dumitru, N. (2013). The Effect of Change of Carburizing Media Nature on Growth Kinetics of Layer. *U.P.B. Sci Bull, Series B. 75*, 181–192.

5. Sharma, C. P. (2004). *Engineering Materials: Properties and Application of Metal and Alloys*, PHI Publication: India.

6. Solberg, J. K., Borvik, T., & Lou, D. C. (2009). Surface Strengthening Using a Self-Protective Diffusion Paste and Its Application for Ballistic Protection of Steel Plates. *Materials and Design, 30*, 3525–3536.

7. Cavaliere, P., Zavarise, G., & Perillo, M. (2009). Modeling of The Carburizing and Nitriding Processes. *J. Computational Materials Science, 46*, 26–35.

8. Chen, Y. C. (1992). Effect of H_2O Content in Air on the Carburizing Behavior of Charcoal Gas. *Journal of Materials Engineering and Performance*, 383–391.

9. Kalpajian, S., & Schmid, S. R. (2010). *Manufacturing Engineering and Technology, Sixth Edition in Si Units*, Prentice Hall: USA.

10. Hassan, M., & Shafiq, M. (2010). Pulsed Ion Beam-Assisted Carburizing of Titanium In Methane Discharger. *Chinese Physical B. 19*, 1–10.

11. Gorockiewicz, R., & Lapinski, A. (2010). Structure of the Carbon Layer Deposited on the Steel Surface After Low Pressure Carburizing. *Vacuum J. 85*, 429–433.

12. Yoon, J. H., Jee, Y. K., & Lee, S. Y. (1999). Plasma Paste Boronizing Treatment of the Stainless Steel AISI 304. *Surface and Coatings Technology, 112*, 71–75.

13. Elzanaty, H. (2014). The Effect of Carburization on Hardness and Wear Properties of the Mild Steel. *International Journal of Innovation and Applied Studies. 6*, 995–1001.

14. Alias, S. K., Abdullah, B., Abdullah, A. H., Latip, S. A., Wahab, N. A., & Ghani, M. A. A. (2013). Mechanical Properties of Paste Carburized ASTM A516 Steel. *Procedia Engineering*, 525–530.

15. Leontin, & Dragomir, D. (2001). The Advantages of Fluidized Bed Carburizing. *Material Science and Engineering*, 115–119.

CHAPTER 11

NEW CYCLIC THIOUREA FOR CORROSION INHIBITION: SYNTHESIS, CHARACTERIZATION AND CORROSION INHIBITION STUDIES OF 1,3-BIS(N'-2,3,4-METHYL BENZOYLTHIOUREIDO)-1,3-METHYLBENZENE

KARIMAH KASSIM, NOOR KHADIJAH MUSTAFA KAMAL, SITI NORIAH MOHD SHOTOR, and ADIBATUL HUSNA FADZIL

Faculty of Applied Sciences, Universiti Teknologi MARA, Shah Alam, Malaysia

CONTENTS

OVERVIEW

A series of bisthiourea ligands, namely 1,3-bis(N'-2-methylbenzoylthioureido)-1,3-methylbenzene(B1), 1,3-bis(N'-3-methylbenzoylthioureido)-1,3-methyl-benzene(B2) and 1,3-bis(N'-4-methylbenzoylthioureido)-1,3-methylbenzene (B3) have been successfully synthesized and characterized by using CHNS elemental analyzer, Fourier Transform Infra Red (FTIR) and ^{1}H and ^{13}C Nuclear Magnetic Resonance (NMR) spectroscopies. The inhibition performance of each compounds were studied using weight loss method and linear polarization resistance (LPR) technique. The treatment was achieved by immersion of mild steel coupons in 1.0 M sulfuric acid solutions with variable concentrations of the compounds ranging from 1×10^{-3} M to 1×10^{-5} M. Results obtained from both technique shows that all the compounds were able to reduce the corrosion rate and compound B3 has the highest inhibition efficiency percent which is in a range of 74.3 to 92.3%.

11.1 INTRODUCTION

Organic compounds have been reported to be the effective corrosion inhibitors for mild steel in sulfuric acid solutions [1–3]. Organic compounds that contain sulfur, nitrogen and oxygen atoms are able to retard metallic corrosion. As thiourea derivatives molecules contains both sulfur and nitrogen atoms, these molecules are the potential corrosion inhibitors [4–6]. These atoms become the center of adsorption as well as multiple bonds that are present in the molecules or aromatic rings as the substituents. The adsorption of electron lone pairs of the donor atoms of the inhibitor with the metal surface will form a film that reduces the corrosive attack in an acid medium [7, 8]. Others investigation states that heterocyclic ring structure which contains nitrogen and oxygen atoms can enhance greater adsorption on metal surface [9]. Therefore, cyclic structure of thiourea is the best organic compound that can be used as the corrosion inhibitor. In this study, bis-thiourea ligands containing ring structure were synthesized and used as corrosion inhibitor. However, there were only few studies were conducted on these type of ligands to inhibit corrosion activity. The aim of this study is to investigate the inhibitory efficiency of bis-thiourea ligands that were synthesized via substitution and addition reaction using *m*-xylylenediamine, ammonium thiocyanate and *o, m, p*-methylbenzoyl chloride as in Figure 11.1.

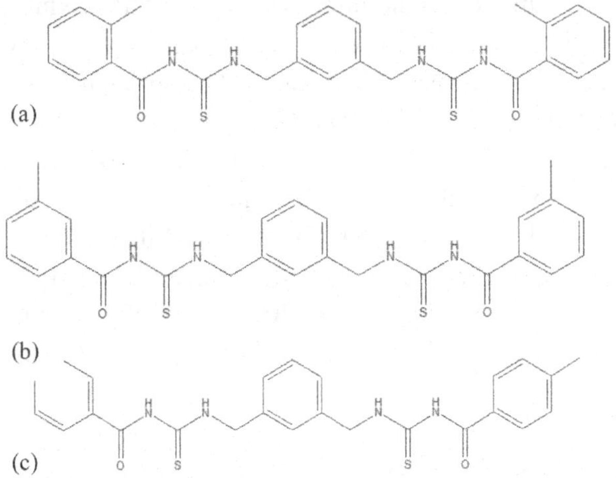

FIGURE 11.1 The chemical structure of the investigated compounds:(a) 1,3-bis (*N'*-2-methylbenzoylthioureido)-1,3-methylbenzene (B1); (b) 1,3-bis(*N'*-3-methylbenzoylthioureido)-1,3-methylbenzene (B2) and (c) 1,3-bis(*N'*-4-methylbenzoylthioureido)-1,3-methylbenzene.

11.2 EXPERIMENTAL

11.2.1 PHYSICAL MEASUREMENT

All chemical and solvents were purchased and used without further purification. Melting points were measured using BÜCHI Melting Point B–545. Infrared spectra were recorded using FTIR Perkin Elmer 100 Spectrophotometer in the spectral range of 4000–400 cm^{-1} by using KBr pellet. ^{1}H and ^{13}C NMR spectra were obtained from Bruker Advance III 300 Spectrometer at room temperature. The elemental analyzes of compound were performed by CHNS Analyzer Flash EA 1112 series.

11.2.2 SYNTHESIS OF BIS-THIOUREA LIGANDS

11.2.2.1   Synthesis of 1,3-Bis(N'-2-Methylbenzoylthioureido)-1,3 Methylbenzene

A 5 mmol of 2-methylbenzoyl chloride was added to acetone solution together with 5 mmol of ammonium thiocyanate. The mixture was stirred for about 15 minutes. A solution of *m*-xylylenediamine (2.5 mmol) in acetone was added and refluxed for 3 hours. The solution was poured into a beaker containing ice cubes. The resulting precipitate was collected by filtration and recrystallized with ethanol. Yield 65.7%; white solid, melting point: 156°C. IR (KBr pellet, cm^{-1}): v(N-H) 3186.88, v(C=O) 1656.38, v(C-N) 1168.06, v(C=S) 739.99. ^{1}H NMR δ 2.30 (s, 3H, CH$_3$); 7.20–7.50 (m, Ar-H); 9.10 (d, H, CONH); 11.15 (s, H, CSNH).^{13}C NMR δ 29 (CH$_3$); 126–161 (aromatic ring); 201 (C=O); 208 (C=S). Anal. Calc. for C$_{26}$H$_{26}$N$_4$O$_2$S$_2$: C, 63.67; H, 5.31; N, 11.43; S, 13.06. Found: C, 66.40; H, 3.53; N, 11.11; S, 10.20

11.2.2.2   Synthesis of 1,3-Bis(N'-3-Methylbenzoylthioureido)-1,3-Methylbenzene

This was prepared as above from equimolar of 3-methylbenzoyl chloride, ammonium thiocyanate and *m*-xylylenediamine. Yield 68.2%; light yellowish solid, melting point: 118°C. IR (KBr pellet, cm^{-1}): v(N-H) 3266.21, v(C=O) 1633.53, v(C-N) 1268.78, v(C=S) 740.82. ^{1}H NMR δ 2.30 (s, 3H, CH$_3$); 7.10–7.50 (m, Ar-H); 9.0 (d, H, CONH); 11.25 (s, H, CSNH).^{13}C NMR δ 26 (CH$_3$); 122–150 (aromatic ring); 212 (C=O); 218 (C=S). Anal. Calc. for C$_{26}$H$_{26}$N$_4$O$_2$S$_2$: C, 63.67; H, 5.31; N, 11.43; S, 13.06. Found: C, 64.66; H, 5.29; N, 11.06; S, 10.01.

11.2.2.3   Synthesis of 1,3-Bis(N'4-Methylbenzoylthioureido)-1,3-Methylbenzene

This was prepared as above from equimolar of 4-methylbenzoyl chloride, ammonium thiocyanate and *m*-xylylenediamine. Yield 63.8%; white solid, melting point: 162°C. IR (KBr pellet, cm⁻¹): v(N-H) 3246.26, v(C=O) 1668.87, v(C-N) 1256.53, v(C=S) 745.27. ^{1}H NMR δ 2.40 (s, 3H, CH$_3$); 7.15–7.35 (m, Ar-H); 9.15 (d, H, CONH); 11.30 (s, H, CSNH).^{13}C NMR δ 32 (CH3); 118–151 (aromatic ring); 195 (C=O); 203 (C=S). Anal. Calc. for C$_{26}$H$_{26}$N$_4$O$_2$S$_2$: C, 63.67; H, 5.31; N, 11.43; S, 13.06. Found: C, 64.02; H, 4.68; N, 11.22; S, 10.50.

11.2.3 WEIGHT LOSS METHOD

The mild steel coupons of size of 2 × 2 cm^2 were used as specimen. The coupons were first abraded with emery paper and then washed with acetone followed by distilled water and dried. The coupon were immersed for a week in 1.0 M H$_2$SO$_4$ solution with and without inhibitors and the concentrations of the inhibitors used are varies from 1 × 10⁻⁵ M, 1 × 10⁻⁴ M to 1 × 10⁻³ M. The weight loss of each coupon was recorded.

11.2.4 LINEAR POLARIZATION RESISTANCE (LPR) TECHNIQUE

This method was carried out in a three-electrode cells using AUTOLAB instrument equipped with NOVA software. A saturated calomel electrode (SCE) was used as a reference electrode and a graphite electrode as a counter. The working electrode was prepared by embedding a rod with mild steel coupon in epoxy resin, with an exposed surface area of 0.065 cm^2. The surfaces were polished with emery paper and washed with distilled water for electrochemical studies.

11.3 RESULTS AND DISCUSSION

11.3.1 FTIR SPECTROSCOPY

All the expected frequency region for the three synthesized compounds such as v(N-H), v(C=O), v(C-N) and v(C=S) had been observed around 3200 cm⁻¹, 1600 cm⁻¹, 1200 cm⁻¹ and 700 cm⁻¹, respectively. The N-H band was observed above 3200 cm⁻¹ which is due to the presence of intramolecular hydrogen bonding. Meanwhile, it is clearly observed that the carbonyl band was found above 1600 cm^{-1}. This is because the resonance effect of the phenyl rings and the existence of intramolecular hydrogen bonding with N-H [10]. The C=S absorption band was observed around 700 cm^{-1} and it in close agreement with the previous study with a higher absorption of the C=S band. The presence of the aromatic group affected the absorption band in which it is describe as large double bond character and has lower nucleophilic character of sulfur atom compared to alkylthioureas. This assignment of bis-thiourea derivatives were confirmed by Halim and his co-workers study [11, 12].

11.3.2 ^1H AND ^{13}C NMR SPECTROSCOPY

The NMR spectra of all compounds for H shift show that there was methyl proton exist in it in a range of 2.30–2.40 ppm and aromatic proton between 7.20–7.50 ppm. There were two resonances for amine group, where the proton near to the carbonyl group and the other one was for the thioamide group. For carbonyl, the resonance was observed between 9.1–9.15 ppm. On the other hand, the proton shifts resonance for thioamide group was observed at 11.15–11.30 ppm. As in theory, the resonance for this type of group should be more than 12 ppm, but in this study, it was observed at lower resonance due to the presence of the aromatic group in the thiourea compounds in which the electrons on the aromatic ring deshield the hydrogen attached to amine group. Apart from that, the chemical shift of this group is variable. It is depends not only on the chemical environment in the molecule but also on the concentration, temperature and solvent. As for the ^{13}C, the NMR spectra were in agreement with the proposed structure. The carbon atoms for CH_3, C=O and C=S can be observed around δ 30 ppm, δ 200 ppm and δ 210 ppm.

11.3.3 WEIGHT LOSS METHOD

The inhibition efficiency percent of each synthesized compounds towards corrosion and its respective corrosion rate had been calculated through the information recorded from weight loss method. The inhibition efficiency percent and the corrosion rate has been calculated by using the Equation 11.1 and Equation 11.2, respectively; and listed as in Table 11.1.

$$IE\% = \left(\frac{W_{blank} - W_{inhibitor}}{W_{blank}} \right) \times 100\% \qquad (11.1)$$

where, W_{blank} is the weight loss of mild steel without inhibitor and $W_{inhibitor}$ is the weight loss of mild steel in with inhibitor.

$$\text{Corrosion rate} \left(g\,cm^{-2}day^{-1} \right) = \frac{\text{Weight loss of mild steel} \left(g \right)}{\text{Surface area of mild steel} \left(cm^{2} \right) \times \text{Time} \left(day \right)} \qquad (11.2)$$

TABLE 11.1 Parameters of Mild Steel Corrosion Activities in 1.0 M H_2SO_4 Solution With and Without Inhibitors

Compound	Concentration (M)	Average weight loss (g)	Inhibition efficiency (IE %)	Corrosion rate (g/cm^{-2} day^{-1})
Blank	–	0.1946	–	6.95×10^{-3}
B1	1×10^{-5}	0.0973	50.0	3.48×10^{-3}
	1×10^{-4}	0.0391	79.9	1.40×10^{-3}
	1×10^{-3}	0.0166	91.5	5.93×10^{-4}
B2	1×10^{-5}	0.0996	48.8	3.56×10^{-3}
	1×10^{-4}	0.0766	60.6	2.74×10^{-3}
	1×10^{-3}	0.0187	90.4	6.68×10^{-4}
B3	1×10^{-5}	0.0829	57.4	2.96×10^{-3}
	1×10^{-4}	0.0353	81.9	1.26×10^{-3}
	1×10^{-3}	0.0149	92.3	5.32×10^{-4}

It is clearly shown that the bis-thiourea compounds shown inhibition activity towards the corrosion attack. This is because the presence of N, S and O atom in all compounds synthesized that acts as the center of adsorption to the mild steel. It forms a protective film on the surface of mild steel via chemisorption process by which it decreases the mild steel area from corrosion attack. It was clearly observed that as the concentration of the

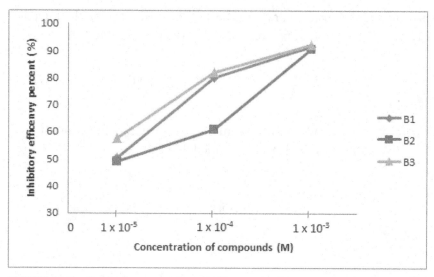

FIGURE 11.2 Graph of concentration of inhibitors versus inhibitory efficiency (IE%) by using weight loss method.

compounds increase, the inhibition efficiency percent of all synthesized compounds also increases. The highest inhibition activity is at concentration of inhibitors at 1×10^{-3} M and the highest inhibition efficiency percent is exhibit by compound B3. The IE% achieved by this compound is 92.3% as seen in Table 11.1 and Figure 11.2. The efficiency of each compounds were affected by the position of methyl substituent on the phenyl ring which are *ortho-*, *meta-* and *para-* positions. This methyl group is an electron donating which it is more favored to *ortho-* and *para-* positions. However, in this study it is most favored to the *para-*position because the methyl group can easily donate the electron to the ring system compare to the *ortho-* position.

11.3.4 LPR TECHNIQUE

Figure 11.3 and Table 11.2 show linear polarization curves and data of mild steel in 1.0 M sulfuric acid with and without inhibitor and its parameter, respectively. The efficiency of inhibitors was calculated by Eq. (11.3):

$$IE\% = \frac{\left(I_{corr} - I_{inh}\right)}{I_{corr}} \times 100\% \tag{11.3}$$

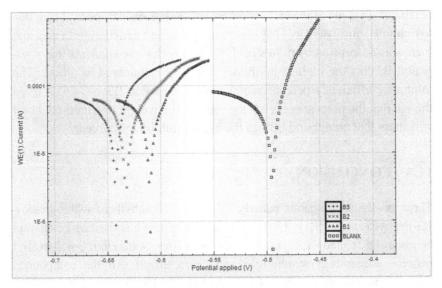

FIGURE 11.3 Polarization curves of 1×10^{-3} M B1, B2 and B3 in 1.0 M sulfuric acid.

TABLE 11.2 Corrosion Parameters for Mild Steel in 1.0 M Sulphuric Acid With and Without Inhibitor

Compound	i_{corr} (μA cm^{-2})	E_{corr} (mV)	bc (V/dec)	ba (V/dec)	IE %
Blank	48.7890	−444.570	28.870	55.820	–
B1	14.0020	−404.010	19.790	30.334	71.3
B2	18.5890	−583.780	66.780	59.940	61.9
B3	12.5550	−380.440	13.423	26.1640	74.3

where I_{corr} and I_{inh} indicate corrosion densities with and without inhibitor, respectively. Other electrochemical parameters such as corrosion density (I_{corr}), corrosion potential (E_{corr}), cathodic and anodic Tafel slopes (bc, ba) and inhibition efficiency are given in Table 11.2.

From Figure 11.3, it shows that the presence of inhibitors caused shifting of the corrosion potential to the negative side as compared to corrosion potential without inhibitor. This means that the reaction occur was a cathodic reaction which lead to hydrogen revolution. Through Table 11.2, it is also observed that the presence of different positions of methyl group lowers the I_{corr} [13] which indicates that the inhibitors suppressed the reactions. This reaction is also affected by the presence of phenyl ring due to rich in electron density [14]. Reducing in current densities indicates that there was adsorption by mild steel [15]. This technique confirms that compound B3 has the highest inhibitory efficiency compared to others. The inhibitory efficiency percent achieved by compound B3 is 74.3%. Even though that the percentage is not the same as the weight loss method, but it still show that compound B3 has the highest inhibitory percent.

11.4   CONCLUSION

Three bis-thiourea ligands namely, 1,3-bis(N'-2-methylbenzoylthioureido)-1,3-methylbenzene (B1), 1,3-bis(N'-3-methylbenzoylthioureido)-1,3-methyl benzene (B2) and 1,3- bis(N'-4-methylbenzoylthioureido) −1,3-methyl benzene successfully synthesized and characterized. All the compounds were able to inhibit corrosion of mild steel in 1.0 M sulfuric acid and the best corrosion inhibitor is compound B3.

ACKNOWLEDGEMENT

The authors are grateful to the Ministry of Higher Education of Malaysia for the research grant no. 600-RMI/RAGS 5/3/(53/2015) and the Faculty of Applied Sciences, Institute of Science and Universiti Teknologi MARA for providing research facilities.

KEYWORDS

- bis-thiourea
- corrosion inhibitor
- ligands
- polarization
- sulfuric acid
- weight loss

REFERENCES

1. Amin, M. A., & Ibrahim, M. M. (2011). Corrosion of mild steel in concentrated H_2SO_4 solutions by a newly synthesized glycine derivatives. *Corrosion Science.* 53, 873–885.
2. Obot, I. B., & Obi-Egbedi, N. O. (2010). Adsorption properties and inhibition of mild steel corrosion in sulfuric acid solution by ketoconazole: experimental and theoretical investigation. *Corrosion Science.* 52, 198–204.
3. Quartarone, G., Bolnado, L., & Tortato, C. (2006). Inhibitive action of indole-5-carboxylic acid towards corrosion of mild steel in deaerated 0.5 M sulfuric acid solutions. *Applied Surface Science.* 252, 8251–8257.
4. Edrah, S., & Hasan, S. K. (2010). Studies on thiourea derivatives as corrosion inhibitor for aluminum in sodium hydroxide solution. *Journal of Applied Science Research.* 6, 1045–1049.
5. Khaled, K. F. (2010). Experimental and molecular dynamic study on the inhibition performance of some nitrogen containing compounds for iron corrosion. *Material Chemistry and Physics.* 124, 760–767.
6. Hosseini, S. M. A., & Azimi, A. (2009). The inhibition of mild steel in acidic medium by 1-methyl-3-pyridin-2-yl-thiourea. *Corrosion Science.* 51, 728–732.
7. Torres, V. V., Rayol, V. A., Magalhães, M., Viana, G. M., Aguiar, L. C. S., Machado, S. P., Orofino, H., & D'Elia, E. (2014). Study of thiourea as derivatives synthesized from a green route as corrosion inhibitors for mild steel in HCl solution. *Corrosion Science*, 79, 108–118.
8. Torres, V. V., Amado, R. S., deSá, C. F., Fernandez, T. L., Riehl, C. A. S., Torres, A. G., & D'Elia, E. (2011). Inhibitory action of aqueous coffee ground extracts on the corrosion of carbon steel in HCl solution. *Corrosion Science.* 53, 2385–2392.
9. Chauhan, J. S., & Gupta, D. K. (2009). Corrosion inhibition of titanium in acidic media containing fluoride with bixin. *E-Journal of Chemistry.* 6, 975–978.
10. Yusof, M. S. M., Jusoh, R. H., Khairul, W. M., & Yamin, B. M. (2010). Synthesis and characterization of *N*-(3,4-dichlorophenyl)-*N'*-(2,3 and 4-methylbenzoyl) thiourea derivatives. *Journal of Molecular Structure*, 975, 280–284.
11. Halim, N. I. M., Kassim, K., Fadzil, A. H., & Yamin, B. M. (2012). Synthesis characterization and antibacterial studies of Cu(II) complexes thiourea. *The Malaysian Journal of Analytical Sciences.* 16(1), 56–61.
12. Halim, N. I. M., Kassim, K., Fadzil, A. H., & Yamin, B. M. (2012). Synthesis characterization and antibacterial studies of 1,2-Bis(*N'*-2,3 and 4-methoxybenzoylthioureido)-4-nitrobenzene. *APCBEE Procedia.* 3, 129–133.
13. Quraishi, M. A., Ansari, F. A., & Jamal, D. (2002). Thiourea derivatives as corrosion inhibitors for mild steel in formic acid. *Materials Chemistry and Physics.* 77, 687–690.
14. Wang, J., Cao, C., Chen, J., Zhang, M., Ye, G., & Lin, H. (1995). Anodic desorption of inhibitors. *J. Chin. Soc. Corros. Protect.* 15, 241–248.
15. Fouda, A. E. E., & Hussein, A. (2012). Role of some phenylthiourea derivatives as corrosion inhibitors for carbon steel in HCl solution. *Journal of the Korean Chemical Society.* 56(2), 264–273.

CHAPTER 12

BIOETHANOL PRODUCTION FROM OIL PALM FROND LIGNOCELLULOSE USING FACULTATIVE ANAEROBIC THERMOPHILIC BACTERIA

SITI NUR RIDHWAH MUHAMED RAMLI,[1]
TENGKU ELIDA TENGKU ZAINAL MULOK,[1]
SABIHA HANIM MOHD SALLEH,[1] KHALILAH ABDUL KHALIL,[1]
OTHMAN AHMAD,[1] NIK ROSLAN NIK ABDUL RASHID,[1] and
WAN ASMA IBRAHIM[2]

[1]*Faculty of Applied Sciences, Universiti Teknologi MARA, Shah Alam, Malaysia*

[2]*Forest Research Institute of Malaysia, Kepong, Malaysia*

CONTENTS

OVERVIEW

This study focused on production of bioethanol from biomass, i.e., oil palm frond lignocelluloses (OPF) using facultative thermophilic anaerobic bacterial strain. Lignocellulose of OPF is comprised of holocellulose, α-cellulose and lignin. The study on characterization was based on OPF treated with steam explosion and OPF without any treatment. The composition of each component was determined by chemical analysis for treated and untreated OPF. The content of holocellulose and α-cellulose for untreated OPF was 57.69% (w/v) and 39.50% (w/v), respectively. The content of holocellulose and α-cellulose showed higher percentage in treated OPF which was 88.05% (w/v) and 46.36% (w/v), respectively. For lignin content, the untreated OPF showed the higher percentage than treated OPF which was 14.28% (w/v) and 10.33% (w/v) for treated OPF. Isolates of facultative anaerobic thermophilic bacteria from hot spring Sungai Klah, Perak, Malaysia were screened for bioethanol production using shake flaks fermentation under the condition at 160 rpm for 120 h and 54°C. The treated OPF was used as the substrate since the results showed high percentage of holocellulose and α-cellulose. All the isolates demonstrated the potential for bioethanol production.

12.1   INTRODUCTION

Bioethanol is produced by fermenting the sugar components in plant materials. It is mostly made from sugar and starch crops. Other bioethanol production using feedstock, with advanced technology being developed is cellulosic biomass such as trees and grasses. In its pure form bioethanol can be used as a fuel for vehicles, but it is commonly used as a gasoline additive to increase octane and improve vehicle emissions. In addition, bioethanol is an important renewable energy source due to the economic and environmental benefits [1]. To replace the non-environmentally friendly fossil hydrocarbons are the raw materials which are renewable and thus creating "green" products. In contrast to traditional fuels, fermentation of bioethanol does not contribute to the green house effect, being a CO_2 neutral resource [2]. It is also widely used in USA and Brazil [3]. It is renewable energy source since it is produced from plants that are grown and harvested every year that can be found or planted in most parts of the world.

Moreover, bioethanol from cellulosic biomass provides a low cost and abundant resource that has the potential to support large scale production of fuels and chemicals [4]. Lignocellulosic biomass includes materials such as agricultural residues, forestry residues, portions of municipal solid waste and various industrial wastes. Herbaceous and woody crops are also included. Lignocellulosic biomass is composed of cellulose, hemicelluloses and lignin [5]. The carbohydrate polymers (cellulose and hemicelluloses) are tightly bound to the lignin which is also the major component of lignocellulosic [6]. The plant cells are composed of different layers, which differ from one another with respect to their structure and chemical composition. Basically, a skeleton is surrounded by other substances serving as matrix (hemicelluloses) and materials in a form of cellulose. Between lignin and polysaccharides (lignin-carbohydrate complexes, LCC), there are closely associated and covalent cross-linkages of cellulose, hemicelluloses and lignin [7].

Furthermore, lignocellulosic materials hydrolyze and fermentation processes are very complicated. Factors influencing the yields of the lignocelluloses to the monomeric sugars and the by-products are particle size, liquid to solid ratio, type and concentration of enzymes used, temperature, reaction time, the length of the macro molecules, degree of polymerization of cellulose, configuration of the cellulose chain, association of cellulose with other protective polymeric structures within the plant cell wall such as lignin, pectin, hemicelluloses, proteins, and mineral elements. Therefore, pretreatment is a process required for the use of lignocellulosic materials to obtain a high degree of sugar fermentation [8].

Biomass structure is affected by pretreatment of solubilizing hemicelluloses, reducing crystallization and increasing the available surface area and pore volume of the substrate. There are several methods of pretreatment or combined pretreatment methods available. Generally, pretreatment techniques can be categorized into three categories: physical, chemical, and biological. Physical pre-treatment methods included comminution, steam explosion, milling and grinding, extrusion and expansion, high pressure steam and others [9]. The cost-effective way of cellulosic biomass pretreatment is the main challenge of the research and development of cellulose ethanol technology.

With hydrolysis and fermentation processes, lignocellulosic biomass can be converted into bioethanol. In hydrolysis, the cellulosic biomass is converted to sugars and fermentation involves converting sugar to bioethanol [10]. Fermentation involves a series of chemical reaction during conversion of biomass simple sugars to bioethanol in the presence of microorganisms such as yeasts and bacteria. The by-product of fermentation is CO_2.

Thus, the main objective of this study is to characterize lignin, cellulose and hemicelluloses of OPF before and after pretreatment obtained by using steam explosion and to isolate potential bioethanol producer from local hot spring using lignocelluloses in OPF as carbon source in Shake Flask Fermentation.

12.2 MATERIALS AND METHODS

12.2.1 SAMPLE COLLECTION AND BACTERIAL ISOLATION

The sample was collected from natural hot spring located in Sungai Klah, Sungkai, Perak, Malaysia. The temperature was 54°C and the pH was 7.6. One ml of water and sediment mixture from hot spring was inoculated into 100 ml of nutrient broth in 250 ml Erlenmeyer flask and shaken (160 rpm) at temperature 54°C for 24–48 h.

12.2.2 CHARACTERIZATION OF OPF

Chemical analysis was used to determine the composition of cellulose, hemicelluloses and holocellulose. Moisture content: The empty crucibles were weighed prior to addition of 2 g of sample in the crucibles. The sample was dried overnight in the oven at 105°C. The drying was repeated until a steady weight was achieved.

12.2.3 ALCOHOL-BENZENE SOLUBILITY

About 2 g of sample was added into a priorly weighed thimble. A mixture of ethanol-toluene (1:1) was added into the round bottom flask (RBF).

The thimbles were assembled in the soxhlet apparatus followed by extraction using ethanol- toluene solution for 8 hours. After extraction, the RBF was dried in the oven at 105°C overnight. The drying process was repeated until a steady weight was achieved.

12.2.3.1   Holocellulose

The extracted residue from ethanol-toluene solubility was transferred into a 250 ml conical flask. The following mixture was later added into the conical flask: (i) 100 ml distilled water (ii) 1.5 g $NaClO_2$ (iii) 5 ml 10% (v/v) acetic acid. The mixture was heated using a boiling water bath. The adding process was continued until a total 6 g of $NaClO_2$ was added. Boiling was continued for another 30 min. The residue obtained was white but still retained the woody structure. After cooling the residue was filtered using pre-weighted fritted glass crucibles (porosity 1). The sample was later washed with ice-distilled water. Finally, the residue was washed with acetone and later air-dried.

12.2.3.2   α-Cellulose

The following experiment was carried out in a water bath at 20°C. The air-dried holocellulose was added in to a beaker containing 15 ml of 17.5% (w/v) NaOH and later macerated for 1 min. Another 10 ml of NaOH was later added and mixed for 45 secs followed by the third addition of 10 ml of NaOH and mixed for 15 sec. The mixture was allowed to stand for 3 min. Finally another 10 ml of NaOH was added and mixed for 2½ min. The third step involving the addition of NaOH were repeated three times. The mixture was allowed to stand for 30 min. One hundred ml of distilled water was later added and left to stand for 30 min. The mixture was filtered using a pre weighed fritted glass crucible (porosity 3). The filtered mixture was washed using 25 ml of 8.3% (w/v) NaOH followed by 650 ml of distilled water (20°C). The crucible was later filled with 2N acetic acid and allowed to stand for another 5 min. Acetic acid was removed using suction followed by distilled water to free the filtrate from any acid. The crucible was then dried at 105°C until a constant weight was achieved.

12.2.4   DETERMINATION OF LIGNIN

The following experiment was carried out in 20°C water. About 1 g of sample from alcohol-benzene solubility extraction and 15 ml 72% H_2SO_4 was added into a beaker enclosed with a lid and allowed to react for 2 hours. The mixture was later stirred before transferring into a 1 L conical flask containing 300 ml distilled water. Distilled water was added until the concentration of H_2SO_4 was reduced from 72% (v/v) to 3% (v/v). The mixture was heated for 4 hours and upon cooling, was filtered using filter glass crucible (FGC) (porosity 4). Then the residue was washed with hot water and dried in the oven (105°C) until constant weight was achieved.

12.2.5   PRETREATMENT

Steam explosion was used for the treatment of OPF for 30 minutes under the following conditions: temperature at 160°C and pressure at 6 bar.

12.2.6   SCREENING FOR ETHANOL PRODUCTION USING SHAKE FLASK FERMENTATION

One ml of enrichment culture was inoculated into 150 ml of minimal salt medium (MSM) in 250 ml Erlenmeyer flask and shaken (160 rpm) at temperature 54°C for 5 days. Samples were serially diluted and plated by spread plate method on nutrient a gar and incubated at temperature 54°C for 24 hours. The composition of the MSM (g/l): Bactopeptone – 5; Glucose – 30; NH_4Cl – 2; KH_2PO_4 – 1; $MgSO_4.7H_2O$ – 0.3. Ethanol content was determined using refractometer kit (Refractometer, PAL–1, Atago).

12.3   RESULTS AND DISCUSSION

12.3.1   CHARACTERIZATION OF OPF/PRETREATMENT

Pretreatment is a process that removed lignin and hemicelluloses from lignocellulosic material. It can enhance the hydrolysis of cellulose, meaning smaller percentage of lignin would enhance the hydrolysis of cellulose. As a result, higher of ethanol would be produced. Lignin content for the treated OPF was 10.33% (w/w) and 14.28% (w/w) for untreated OPF as shown in Figure 12.1 indicating that treated OPF is a better raw material that can produce higher percentage of ethanol.

High lignin content of OPF blocks enzyme accessibility, causing end product inhibition, and reducing the rate and yield of hydrolysis. In addition to lignin, cellobiose and glucose can also act as strong inhibitors for celluloses [11]. The carbohydrate in plant is composed of cellulose and hemicelluloses polymers with minor amounts of other sugar polymers such as starch and pectin. In this study, the holocellulose (α-cellulose and hemicelluloses) content in the sample was determined by delignification or lignin removal using bleaching agent and acid. The average holocellulose content in the untreated OPF was 57.69% (w/w) whilst 88.05% (w/w) was obtained from treated OPF, again indicating that application of treatment to the biomass materials would increase the holocellulose content. OPF treated with steam explosion had lower content of ethanol-toluene extractable which was 2.89% (v/v) compared to untreated OPF which was 3.40% (v/v). There as on why the OPF had lower content of unwanted compounds could be due to during the pretreatment process, all the unwanted

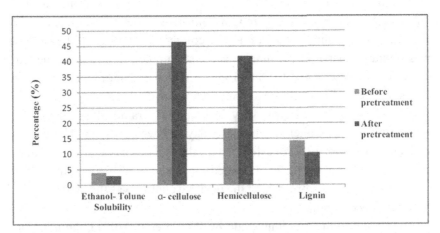

FIGURE 12.1   Percentage composition of holocellulose, α-cellulose, hemicellulose and lignin in OPF and ethanol-toluene solubility before and after pretreatment of the OPF.

compounds were already removed. The α-cellulose content of treated OPF was 46.36% (w/w) and in untreated OPF was 39.50% (w/w) where the reaction involved hydrolysis of amorphous hemicelluloses using diluted acid. α-cellulose is resistant to mild hydrolysis due to its crystalline structure and hence remained as residue in the process [12]. The lignin, cellulose the hemicelluloses contents of the treated and untreated OPF in this study were similar to those obtained by previous researchers Goh et al. (2010) which were ((12.96% (w/w), 30.19% (w/w) and 24.26% (w/w)) and [(19.23% (w/w), 44.78% (w/w) and 35.99% (w/w)], respectively. It has been reported that the pretreatment used in their study were through hydrothermal treatment followed by enzymatic hydrolysis where the OPF was prepared by treating the pre-hydrolyzed biomass using 3% (v/v) sulfuric acid. In this study, there was also an increment in cellulose and hemicelluloses content after the pretreatment of the OPF using physical method, i.e., stem explosion although it did not involve any chemical or pre-hydrolysis using sulfuric acid.

12.3.2 BACTERIAL ISOLATION AND SCREENING FOR BIOETHANOL PRODUCTION

A total of 9 isolates were collected from hot spring under thermophilic condition. The isolates were characterized morphologically as indicated in Table 12.1. All the isolates were Gram positive, except for isolates B and F, were all rod in shape and motile. The colony morphology of the isolates was round in shape, entire margin and rise in colony elevation. The isolates were later screened for the production of bioethanol when grown in minimal salt medium (MSM) at temperature 54°C.

The samples were drawn every 24 h and later rapid screened for ethanol using refractometer kit during 5 days of fermentation. The results indicated in Figure 12.2 revealed that all the isolates have the potential of being a bioethanol producer. The medium without any inoculums added served as negative control whilst the positive control was the medium with *Saccharomycescereviciae*.

According to the previous study by Prasad et al. (2007), the microorganisms used in the fermentation for bioethanol production are normally the fungi *Trichodermareesei* and *S. cerevisiae*. They also found that the

TABLE 12.1 Characterization of the Isolates Based on Gram Staining, Shape, Motility and Colony Morphology

Isolate	Gram staining	Shape	Motility	Colony
A	P	Rod	M	Rise, entire, round
B	N	Rod	M	Rise, entire, round
C	P	Rod	M	Rise, entire, round
D	P	Rod	M	Rise, entire, round
E	N	Rod	M	Rise, entire, round
F	P	Rod	M	Rise, entire, round
G	P	Rod	M	Rise, entire, round
H	P	Rod	M	Rise, entire, round
I	P	Rod	M	Rise, entire, round

Note: P (Gram-positive); N (Gram-negative); M (Motile).

thermotolerant yeast *Candida acidothermophilum*, produced 80% (v/v) of the theoretical ethanol yield at 40°C. The thermo tolerant bacterial species in this study, also having the potential of being a bioethanol producer, is more advantageous compared to fungus since their growth rate is substantially higher than fungus.

During fermentation, there were two steps involving different enzymatic systems responsible for the conversion of cellulose into glucose as shown in Scheme 12.1. During the first step (Step 1), the enzyme β-1, 4 glucanase cleaved the glucosidic linkage in cellulose forming cellobiose, which is a glucose dimer with a β-1, 4 bonds as opposed to maltose, a counter part with an α-1, 4 bonds. Subsequently, this β-1,4 glucosidic linkages is broken by the enzyme β-glucosidase forming glucose in step 2 [13].

Various species of fungi and bacteria produce cellulose enzyme and transported across the cell membrane to the external environment. It is common to refer to a mixture of compounds that can degrade cellulose

$$\text{Cellulose} \xrightarrow[\text{Step 1}]{\beta-1,\ 4\text{glucanase}} > \text{Cellobiose} \xrightarrow[\text{Step 2}]{\beta-\text{glucosidase}} > \text{Glucose}$$

SCHEME 12.1 Fermentation of cellulose into glucose.

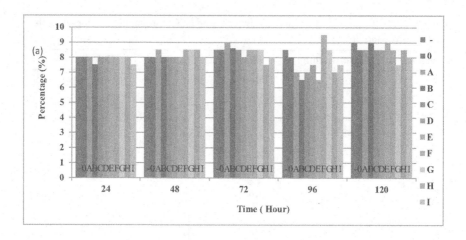

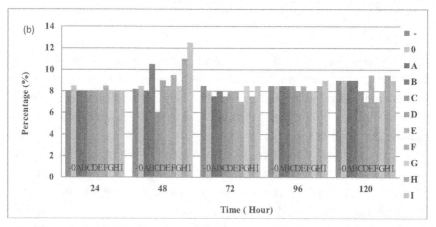

FIGURE 12.2   Ethanol production (%) during shake flask fermentation for 120 h (a) before and (b) after pretreatment. *Note*: 1 (Negative control), 0 (Positive control).

as cellulase, since it is comprised of more than one distinctive enzymes [14]. Recent finding has shown that one of the relatively inert components of the enzyme has the ability to recognize and attach itself to the surface of the cellulose mass, in addition to the ability of recognizing and holding the other protein components that exhibit enzymatic activities [15]. Therefore, the chance of response is even more enhanced by the effect of distance, because the active enzyme is held onto the surface of a solid substrate by an inert protein acting as glue. In general, different species of microorganisms produce different cellulolytic enzymes. *Saccharomyces cerevisiae,* known as baker's yeast, was used as positive controls since it is commonly used fermentation yeast [5]. It is also the most commonly used microorganism for bioethanol production due to its ability to grow in high sugar concentration and producing high yield of bioethanol. The ethanol production after pretreatment as shown in Figure 12.2(b) was high compared to before pretreatment (Figure 12.2 (a)). The isolated bacterial strain also represented a promising candidate for cellulose production in addition to them being thermophilic (less contamination problem and faster growth rate at high temperature), and an aerobic (no oxygen transfer limitation).

12.4   CONCLUSION

Among the various renewable sources of energy, biomass stands out as one of the most promising. Besides its tremendous quantity, there are many advantages of using plant biomass as a source of energy. In addition to the application of the OPF as a substrate for ethanol production, it would also solve the problem of waste disposal which can be very costly. Another factor that should be taken into consideration in reducing the downstream cost is the treatment of the OPF prior to fermentation which can affect the yield of ethanol production as demonstrated in this study. The pretreatment of the OPF using steam explosion showed an increment in cellulose and hemicelluloses content that would obviously increase the yield of ethanol. The composition of each component was determined by chemical analysis for treated and untreated OPF. The content of holocellulose and α-cellulose for untreated OPF was 57.69% (w/w) and 39.50% (w/w), respectively. The content of holocellulose and α-cellulose showed higher percentage in treated OPF which was 88.05% (w/w) and 46.36% (w/w), respectively. For lignin content, the untreated OPF showed higher percentage compared to treated OPF which was 14.28% (w/w) and 10.33% (w/w) for treated OPF. The pretreatment method was cost-effective and sustainable to support the biotechnology industries in the long term. Also in these findings, potential lignocelluloses degradable facultative anaerobic thermophilic bacteria were isolated from local hot spring.

ACKNOWLEDGEMENT

The authors wish to acknowledge the help of Forest Research Institute Malaysia, individuals whose literature is cited in this article and Research Management Institute (RMI) Universiti Teknologi MARA (UiTM) for funding this project (600-RMI/DANA 5/3/RIF (563/2012).

KEYWORDS

- **bioethanol**
- **facultative anaerobic thermophilic bacteria**
- **oil palm frond**
- **response surface methodology**

REFERENCES

1. Sveinsdottir, M., Rafn, S., Baldursson, B., & Orlygsson, J. (2009). Ethanol production from monosugars and lignocellulosic biomass by thermophilic bacteria isolated from Icelandic hot springs. Iceland, *Agriculture Science, 22*, 45–58.
2. Millati, R., Niklasson, C., & Taherzadeh, M. J. (2002). Effect of pH, time and temperature of over-liming on detoxification of dilute-acid hydrolyzates for fermentation by *S. cerevisiae*. *Process Biochem, 38*, 515–22.
3. Kim, S., & Dale, B. E. (2004). Global potential bioethanol production from wasted crops and crop residues. *Biomass and Bioenergy, 26*, 361–375.
4. Sarkar, N., Ghosh, S. K., Bannerjee, S., & Aikat, S. (2012). Bioethanol production from agricultural wastes: An overview. *Renewable Energy, 37*, 19–27.
5. Girio, F. M., Fonseca, C., Carvalheiro, F., DuarteL. C., Marques, S., & Bogel-Lukasik, R. (2010). Hemicelluloses for fuel ethanol. *Bioresources Technology, 101*, 4775–4800.
6. Yong, T. L. K., Lee, K. T., Mohamed, A. R., & Bhatia, S., (2007). Potential of hydrogen from oil palm biomass as a source of renewable energy worldwide, *Energy Policy, 35*, 5692–5701.
7. Lu, X., & Saka, S., (2012). Hydrolysis of Japanese beech by batch and semi-flow water under subcritical temperatures and pressures. *Biomass and Bioenergy, 34*, 1089–1097.
8. Kaparaju, P., & Felby, C., (2010). Characterization of lignin during oxidative and hydrothermal pre-treatment processes of wheat straw and corn Stover. *Bioresource Technology, 101*, 3175–3181.

9. Demirbas, A. (2005). Hydrogen production via pyrolytic degradation of agricultural residues, *Energy Sources, 27*(8), 769–775.
10. Zahari, M. A. K. M., Zakaria, M. R., Ariffin, H., Mokhtar, M. N., Salihon, J., Shirai, Y., & Hassan, M. A., (2012). Renewable sugars from oil palm frond juice as an alternative novel fermentation feedstock for value-added products, *Bioresource Technology, 110*, 566–571.
11. Lynd, L. R., Weimer, P. J., van Zyl, W. H., & Pretorius, I. S., (2002). Microbial cellulose utilization: fundamentals and biotechnology, *Microbial Molecular Biology, 66*, 506–577.
12. Wyman, C. E., Decker, S. R., Himmel, M. E., Brady, J. W., Skopec, C. E., & Viikari, L., (2005). Hydrolysis of Cellulose and Hemicellulose, *Polysaccharides, Structural Diversity and Functional Versatility, 2*, 42.
13. Ghose, T. K. (1977). Cellulase biosynthesis and hydrolysis of cellulosic substances, *Advances in Biochemical Engineering, 6*, 25.

CHAPTER 13

JET-SWIRL INJECTOR SPRAY CHARACTERISTICS IN COMBUSTION WASTE OF A LIQUID PROPELLANT ROCKET THRUST CHAMBER

ZULKIFLI ABDUL GHAFFAR, SALMIAH KASOLANG, and AHMAD HUSSEIN ABDUL HAMID

Faculty of Mechanical Engineering, Universiti Teknologi MARA Shah Alam, Selangor, Malaysia

CONTENTS

OVERVIEW

A liquid propellant rocket involves the combustion process in generating the thrust for the purpose of propelling the rocket. However, this process causes the emission of heat and wastes which is harmful to the environment and health. An injector discharging fine droplets size facilitates the combustion process and capable to increase the combustion efficiency. Besides the droplets size, other important spray characteristics of an injector is the spray angle. A larger spray angle increases the exposure of the droplets to the surrounding air or gas, which improves the rates of heat and mass transfer. Jet-swirl injector is one of the injector types capable of producing both fine droplets size and wide spray angle. In real applications, the fineness of the droplets size and the wideness of the spray angle are predominantly dependent on the operating conditions of the system and the injector geometrical design. In this chapter, a fundamental investigation into the characterization of the jet-swirl spray behavior has been discussed. The effect of liquid injection pressure, swirl chamber diameter and flow regimes on the characteristics of spray angle are presented.

13.1 INTRODUCTION

Generally, there exist two types of rocket which are liquid propellant rocket and solid propellant rocket. The liquid propellant rocket is a rocket with the propellants is stored separately as liquids and are injected to the thrust chamber [1]. Liquid propellant rocket are generally used for large rockets such as space launch vehicles and ballistic missiles. Solid propellant rocket are much lighter so they are used in smaller missiles such as air-launched and shoulder-launched missiles. Schematic of one type of liquid propellant rocket (liquid bipropellant rocket) is shown in Figure 13.1.

The thrust chamber or thruster is the combustion device of a liquid propellant rocket where the liquid propellants are metered, injected, mixed, and burned to form hot gaseous reaction products. A thrust chamber consists of three major parts: an injector, a combustion chamber, and a nozzle. Figure 13.2 shows the major parts of a lab-scale thrust chamber designed by Hamid et al. [2].

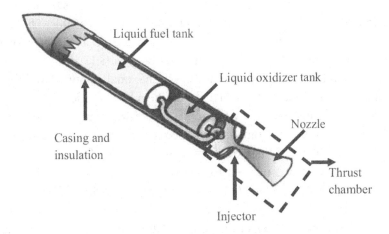

FIGURE 13.1 Schematic of liquid bipropellant rocket (Adapted from [1]).

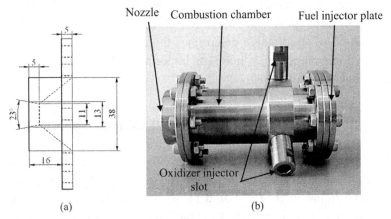

FIGURE 13.2 (a) Nozzle schematic and (b) parts assembled of lab-scale thrust chamber (Adapted from [2]).

In the thrust chamber, the propellants react to form hot gases, which in turn are accelerated and ejected at a high velocity through a supersonic nozzle and hence, imparting momentum to the vehicle. A nozzle was built with a converging section, a constriction or throat, and a conical or bell-shaped diverging section [3].

One of the major public concerns of engine combustion is the pollutant emissions due to their impact on the environment and health. The most common exhaust emissions from liquid rocket engine is carbon dioxide (CO_2), water vapor (H_2O) and alumina/carbon particle which accounting 80% of all rocket exhaust emissions. The smaller portion of liquid rocket exhaust include NO_x, N_2, HCl and others [4]. In order to reduce the emissions of the by-products, an efficient combustion process is required. Particularly, an efficient combustion process in a liquid propellant rocket thrust chamber could be achieved by using a high performance injector. An injector is a device utilized for the process of breaking up bulk liquids into accumulated droplets known as spray. Fine droplets have large surface areas which result in larger exposure of the droplets to the combustion. Hence, a high performance rocket injector refers to the injector which capable of producing liquid propellant sprays with fine droplets. In return, this injector led to an efficient combustion process. An example of a high performance rocket injector is the jet-swirl injector. This injector not only produces sprays with fine droplets but with a large spray angle. The wideness of spray angle determines the exposure of the droplets to the surrounding air or gas, which is an important feature of improving the rates of heat and mass transfer.

13.2   INJECTOR GEOMETRIES AND OPERATING PRINCIPLES

This jet-swirl injector consists of liquid inlet, swirl-generating vanes, swirl chamber, and discharge orifice. Liquid is supplied to the injector through the liquid inlet. The liquid passes through the swirl-generating vanes which create swirling effects on the liquid before exiting the injector through the discharge orifice. The injector is a swirl effervescent injector but in the present study, gas was not introduced into the injector with the purpose of converting the injector into a jet-swirl injector. The schematic of the swirl effervescent injector which act as a jet-swirl injector is shown in Figure 13.3.

The absence of gas causes two of the geometries (the aeration tube and gas inlet) become insignificant features. The reason of converting the swirl effervescent injector to jet-swirl injector is to conduct a comparative analysis of performance between jet-swirl injector and swirl effervescent

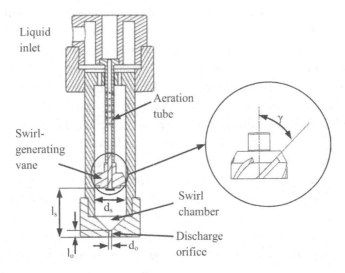

FIGURE 13.3 Schematic of swirl effervescent injector (acts as jet-swirl injector without the introduction of gas).

injector in the future. In jet-swirl injector, swirling insert is deployed to create swirling effects of the liquid inside the injector. The swirl motion of the liquid pushes the flow close to the wall to create a zone of low pressure along the center line. This resulted in air back flow in the injector and hence the formation of air-cored vortex is established. The convergent section near the discharge orifice accelerates the flow prior to exiting the injector [5]. The liquid spreads out in the form of a conical sheet as soon as it leaves the injector and a hollow cone or a solid cone spray is formed due to the breakup of the sheet.

13.3 JET-SWIRL INJECTOR PERFORMANCE PARAMETERS

The performance of an injector is evaluated through spray characterization. Spray characterization is the process of describing a spray. Spray angle is one of the important spray characteristics of a jet-swirl injector besides droplet size as wide spray angle represent wider spray area and better droplets distribution. In combustion process, an increase in spray angle increases the exposure of the droplets to the surrounding air or gas, leading to improved and higher rates of heat and mass transfer [6]. Spray angle is defined as the plane angle formed by the profile of a spray pattern [7]. Figure 13.4 shows the sample image of a spray angle.

Sprays characteristics are influenced by the injector geometries, liquid properties, operating parameters or ambient conditions. Ibrahim [5] mentioned that the performance of the injector is controlled by the liquid properties, injection flow conditions and injector geometry. Particularly, spray angle of swirl injector is dependent on liquid injection pressure and injector geometries. Hamid et al. [9, 10] and Laryea and No. [11] have concluded that the increased of injection pressure widens the spray angle. Rashid et al. [12] have investigated the effect of tangential inlet number on spray angle and observed that, increment of the tangential inlet port number leads to wider spray angle. Hussein et al. [13] have found that injector with larger orifice diameter produces sprays with wider spray angle. Ghaffar et al. [14] have made a review and concluded that most researchers found swirling effect to a liquid stream influences the production of wider spray angle.

Another significant parameter which affects the spray angle in jet-swirl injector is the swirl chamber diameter. Researches have been conducted on the effect of swirl chamber diameter variations towards the spray angle and one of them are conducted by Rizk and Lefebvre [15]. However, the available research conducted the effect of swirl chamber diameter on a pressure swirl injector but not jet-swirl injector. Jet-swirl injector is categorized as a

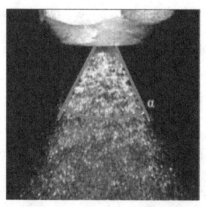

FIGURE 13.4 Image of spray angle, α of a resultant spray produced from an injector (Adapted from Ref. [8]).

different injector from pressure swirl injector by Bayvel and Orzechowski [16]. They stated that jet-swirl atomizer is a combination of jet and swirl atomizer. They also suggested that there exists division of the fluid flow in the jet-swirl atomizer into unswirled axial jet and swirl annular jet due to the nature of the combination. Another lack in the literature is the relation of flow regime to the spray angle characteristics of swirl-related injector.

The main aim of this chapter is to study the effect of injection pressure, swirl chamber diameter and flow regime on the spray angle of a jet-swirl injector. Table 13.1 listed the geometrical parameters of the tested injectors.

TABLE 13.1   Geometrical Parameters of Injectors

Injector no.	Geometrical parameters				
	Vane tip angle, γ (°)	Swirl chamber diameter, d_s (mm)	Swirl chamber length, l_s (mm)	Discharge orifice diameter, d_o (mm)	Discharge orifice length, l_o (mm)
1	45	15	40	2.5	5
2	45	25	40	2.5	5
3	45	40	40	2.5	5

13.4   COLD FLOW TEST

The cold flow test is an alternative method for static firing tests in investigating injector performance. In liquid rocket applications, the liquid propellant is substituted with simulant fluid such as water or other liquids [2, 17–20]. The advantage of this test is it provides clear visualization on atomization and break up processes without the optical distortions initiated by the combustion process [20]. In addition, without the combustion process, this test is safe to the environment. An experimental test-rig was built to perform the cold flow test. A line diagram of the test-rig is shown in Figure 13.5.

A centrifugal pump delivers water from the water supply tank through the water pipeline. Ball valve is installed at the pump outlet to control the amount of water flowing out of the pump. Measurement of water flow rate in the systems is obtained through water flow meter and the flow is

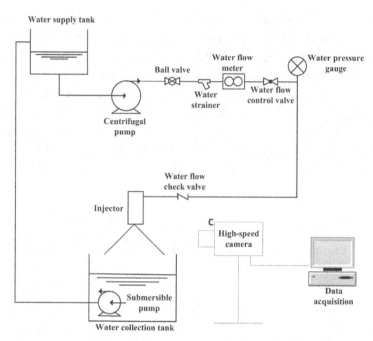

FIGURE 13.5   Line diagram of cold flow test-rig.

controlled by a globe valve. Water strainer is installed at the inlet of water flow meter to prevent unwanted debris pass through the meter which could cause malfunctioned [21]. Water injection pressure is measured by a digital pressure gauge. Water flow check valve installed at the inlet of the injector to prevent backflow especially when both liquid and gas inserted into the injector simultaneously. The injector fixed in vertical downward position produce water sprays into a water collection tank. A submersible pump delivers the water back into the water supply tank to complete the cycle. The cold flow tests of both injectors were performed in the injection pressure range between 0.5 bar and 3 bar with volumetric flow rate between 2 l per-min and 6 l per-min.

13.5   DATA ACQUISITION AND ANALYSIS

The video recordings of the resultant sprays produced were captured by a high-speed video camera with resolutions of 800 x 600 pixels at 1000 frames per second. A halogen flood light is pointed to the sprays to enhance the existing lighting provides by the fluorescent lamp. This is because at high frame rates the image displays on the camera becomes a darker which in need of a brighter light source. Acquired videos were converted to set of images via the video capture software. Image processing software was utilized for further image analysis. Using the image processing software, the edge detection has been performed to separate the spray image from the background. In order to perform the angle measurement process, images were converted to binary to improve the visualization of the spray. Figure 13.6 illustrates the pre-processed and post-processed spray image. Sample measurement of the spray angle is shown in Figure 13.7. The measurement was performed using the angle tool. Intended measurement is highlighted in red.

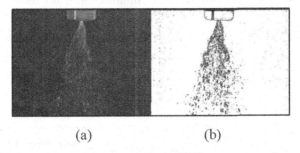

(a)   (b)

FIGURE 13.6   Illustration of (a) pre-processed and (b) post-processed spray image.

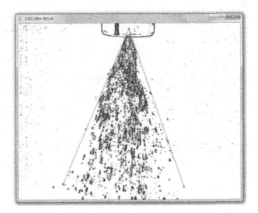

FIGURE 13.7   Sample measurement of spray angle.

13.6   RESULTS AND DISCUSSION

The effect of injection pressure and swirl chamber diameters on spray angle is illustrated in Figure 13.8. It is shown that with $d_s = 15$ mm, the spray angle increase from 40.56° to 54.194° with the increase of the injection pressure from 0.5 to 1.5 bar. Increment of injection pressure beyond this point causes the spray angle to slightly drop to 53.763°. Further increase of injection pressure to 2.5 bar widens the spray angle to 55.5° and no change of spray angle was observed between injection pressure 2.5 bar and 3 bar. The spray angle of $d_s = 25$ mm increase with the increase of injection pressure initially, but decrease afterwards upon exceeding injection pressure 1.5 bar. Further increase in injection pressure up to 3 bar resulted in a steady spray angle starts at injection pressure 2 bar. For $d_s = 40$ mm, the spray angle was observed to be almost unaffected by the increase of injection pressure.

Wider spray angle with increasing liquid injection pressure values is in agreement with previous studies performed by Hamid and Atan [9] and Laryea and No [11]. The reason is, increase of injection pressure increases

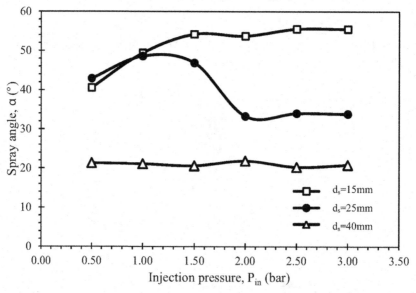

FIGURE 13.8   Effect of injection pressure and swirl chamber diameters on spray angle.

centrifugal force of the swirling liquid. After discharging the orifice, the centrifugal force produced by the rotating liquid sheet converted to the axial velocity component (Skobelkin effect) [22].

However, the existence of abrupt spray angle reduction of $d_s = 25$ mm with increment of injection pressure from 1.5 bar to 2 bar is unexpected. This type of phenomenon was reported to occur only in solid cone injector by Hamid and Atan [9]. The probable reason of this occurrence is the difference of flow regime inside the injector during the injection pressure increment. The flow regime inside all of the injectors with respect to injection pressure is shown in Figure 13.9. It is observed in Figure 13.9 that the flow regime of $d_s = 25$ mm changed from transition flow regime to turbulent flow regime as the injection pressure increase from 1.5 bar to 2 bar. The amount of injection pressure during the transformation of flow regime from transition flow to turbulent flow is high enough to reduce the centrifugal force experienced by the liquid. A smaller centrifugal force resulted in a conversion to a lower axial velocity upon discharging and hence, resulted in a narrower spray angle. However, the reduction of spray angle does not occurs while $d_s = 15$ mm although it experienced the same conversion of transition to turbulent flow regime during the increase of

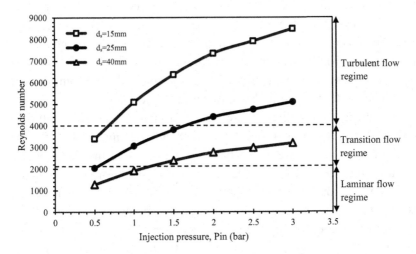

FIGURE 13.9 Relation between injection pressure and Reynolds number for different swirl chamber diameters.

injection pressure from 0.5 bar to 1 bar. The reason is the turbulent flow regime at this condition was not caused by a high injection pressure, but rather a smaller swirl chamber diameter.

It is also observed in Figure 13.8 that a steady spray angle occurred while $d_s = 15$ mm and $d_s = 25$ mm with the increase of injection pressure from 2 bar to 3 bar. This was attributed to the unchanged flow regime (maintained in the turbulent flow regime) experienced by both injectors.

The steady spray angle of $d_s = 40$ mm with the increase of injection pressure from 0.5 bar up to 3 bar as shown in Figure 13.8 is attributed to the flow regime inside the injector. The swirl chamber diameter is too large, hence the flow regime did not achieve a turbulent flow regime with the range of injection pressures as observed in Figure 13.9.

13.7   CONCLUSION

Jet-swirl injectors with a different swirl chamber diameter have been constructed on the effort of investigating the influence of the injection pressures and swirl chamber diameter towards the spray angle. Based on the study, it was found that the increment of injection pressure and reduction of swirl chamber diameter led to the production of a wider spray angle. It is also found that the spray angle depends on the flow regime inside the injectors which determined by the Reynolds number. Flow transformation from transition flow regime to turbulent flow regime with an injection pressure (1.5 bar to 2 bar) of injector 2 ($d_s = 25$ mm) caused the spray angle to drop abruptly although the injection pressure increase. However, the transformation of flow regime from transition flow to turbulent flow of the injector with smaller swirl chamber diameter ($d_s = 15$ mm) at a lower injection pressure (0.5 bar to 1 bar) does not cause the spray angle to reduce. This phenomenon attributed to the combination of the transformation from transition to turbulent flow and sufficient injection pressure of the injector with $d_s = 25$ mm. It is also could be concluded that, despite of only investigating the individual parameter effect on the spray angle, a combine effect or interaction effect between parameters is also crucial to be understood.

ACKNOWLEDGMENT

The authors would like to thank the Ministry of Education of Malaysia (MOE) for the financial support extended to this study through the MyBrain, FRGS and ERGS Grant awards. The authors are also in debt to Research Management Institute of Universiti Teknologi MARA Shah Alam, Malaysia for facilitating this project.

KEYWORDS

- cold flow test
- flow regimes
- jet-swirl injector
- liquid propellant
- spray angle
- swirl chamber

REFERENCES

1. Ward, T. A. (2010). *Aerospace Propulsion Systems.* John Wiley & Sons.
2. Hamid, A. H. A. (2008). Investigation of Swirl Injectors and Supersonic Nozzles for the Development of a Meso-Scale Thrust Chamber. MSc Thesis, Universiti Teknologi MARA.
3. Sutton, G. P., & Biblarz, O. (2001). *Rocket Propulsion Elements, 7th ed.*, John Wiley & Sons: Canada.
4. Ross, M. N., & Sheaffer, P. M. (2014). Radiative Forcing Caused by Rocket Engine Emissions. *Earth's Future, 2*, 177–196.
5. Ibrahim, A. (2006). Comprehensive Study of Internal Flow Field and Linear and Nonlinear Instability of an Annular Liquid Sheet Emanating from an Atomizer. PhD Dissertation, University of Cincinnati.
6. Lefebvre, A. H., Ballal, D. R., & Bahr, D. W. (2010). *Gas Turbine Combustion: Alternative Fuels and Emissions*, CRC Press: Boca Raton, Florida, USA.
7. ASTM, Standard E 1620-97: Standard Terminology Relating to Liquid Particles and Atomization; Standard E 1620-97. (2012).
8. Tratnig, A., & Brenn, G. (2010). Drop Size Spectra in Sprays From Pressure-Swirl Atomizers. *International Journal of Multiphase Flow, 36*, 349–363.

9. Hamid, A. H. A., & Atan, R. (2009). Spray Characteristics of Jet-Swirl Nozzles For Thrust Chamber Injector. *Aerospace Science and Technology*, *13*, 192–196.

10. Hamid, A. H. A., Atan, R., Noh, M. H. M., & Rashid, H. (2011). Spray Cone Angle and Air Core Diameter of Hollow Cone Swirl Rocket Injector. *IIUM Engineering Journal, Special Issue, Mechanical Engineering*, *12*.

11. Laryea, G. N., & No, S. Y. (2004). Spray Angle and Breakup Length of Charge-Injected Electrostatic Pressure-Swirl Nozzle. *Journal of Electrostatics*, *60*, 37–47.

12. Rashid, M. S. F. M., Hamid, A. H. A., Sheng, O. C., & Ghaffar, Z. A. (2012). Effect of Inlet Slot Number on the Spray Cone Angle and Discharge Coefficient of Swirl Atomizer. *Procedia Engineering*, *41*, 1781–1786.

13. Hussein, A., Hafiz, M., Rashid, H., Halim, A., Wisnoe, W., & Kasolang, S. (2012). Characteristics of Hollow Cone Swirl Spray at Various Nozzle Orifice Diameters. *Journal Teknologi*. *58*.

14. Ghaffar, Z. A., Hamid, A. H. A., & Rashid, M. S. F. M. (2012). Spray Characteristics of Swirl Effervescent Injector in Rocket Application: A Review. *Applied Mechanics and Materials*, *225*, 423–428.

15. Rizk, N. K., & Lefebvre, A. H. (1987). Prediction of Velocity Coefficient and Spray Cone Angle for Simplex Swirl Atomizers. *International Journal of Turbo and Jet Engines*. *4*, 65–73.

16. Bayvel, L., & Orzechowski, Z. (1993). *Liquid Atomization*, Taylor & Francis: Washington, DC, USA.

17. Kenny, R. J., Hulka, J. R., Moser, M. D., & Rhys, N. O. (2009). Effect of Chamber Backpressure on Swirl Injector Fluid Mechanics. *Journal of Propulsion and Power*, *25*, 902–913.

18. Soltani, M. R., Ghorbanian, K., Ashjaee, M., & Morad, M. R. (2005). Spray Characteristics of a Liquid–Liquid Coaxial Swirl Atomizer at Different Mass Flow Rates. *Aerospace Science and Technology*. *9*, 592–604.

19. Fu, Q. F., Yang, L. J., & Wang, X. D. (2010). Theoretical and Experimental Study of the Dynamics of a Liquid Swirl Injector. *Journal of Propulsion and Power*, *26*, 94–101.

20. Mayer, W. O. H., Schik, A. H. A., Vielle, B., Chauveau, C., & Gogravel, I. (1998). Atomization and Breakup of Cryogenic Propellants Under High-Pressure Subcritical and Supercritical Conditions. *Journal of Propulsion and Power*, *14*, 835–842.

21. Hedland, M. (2008). *Flow Transmitter Installations and Programming Instructions*. HLIT 306.

22. Yang, V., Habiballah, M., Hulka, J. R., & Popp, M. *Liquid Rocket Thrust Chambers: Aspects of Modeling, Analysis, and Design, vol. 200*, American Institute of Aeronautics and Astronautics (AIAA). (2004).

CHAPTER 14

WEAKNESS OF AUSTENITIC STAINLESS STEEL WELDED JOINTS WHEN SUBJECTED TO HYDROGEN ENVIRONMENT

RASHID DERAMAN and MOHAMAD NOR BERHAN

Faculty of Mechanical Engineering, Universiti Teknologi MARA, Shah Alam, Selangor, Malaysia

CONTENTS

OVERVIEW

Stainless steel is widely used in industrial applications because of its properties are very excellent against corrosion, strength, ductility and weldability. Ferrous metals which bondings by temporary joints are normally be vulnerable to crevice corrosion, thereby causing a welding process to become a first choice to customers. Gas metal arc welding (GMAW) process is a very economical because it has a higher speed and deposition rate with minimal cleanup after welding. Hydrogen can enter the weldment from the parent metal being welded, filler metals and the environment. In this study, a plate of austenitic stainless steel with a thickness of 5 mm was welded using GMAW processes with different welding parameters. A tensile test specimen in standard rectangular shape was subjected to hydrogen charging by immersed, in 0.3 M sulfuric acid solution with current density of 110 mA/cm^2 for 12 hours. Effect of hydrogen environment on the properties and microstructure of the fracture surface of the welding joints was examined using an optical microscope and Phenom XL Desktop scanning electron microscopy (SEM). Experimental results show that the presence of globular oxides on the fracture surface of weldment for the specimen be charged with hydrogen showed that the hydrogen embrittlement phenomenon has a significant influence on the mechanical properties of austenitic stainless steels.

14.1 INTRODUCTION

Stainless steel is widely used in industrial applications because it has features of an excellent combination of corrosion resistance, toughness, ductility and weldability. Henkel and Pense [1] mentioned that this effect achieved by alloying primarily with chromium but may also be enhanced by adding other elements such as nickel and molybdenum. When chromium is added more than about 10% to ordinary steel, extremely thin oxide layer on the surface is transformed and effectively acts as a protective layer in a wide range of corrosive media. George [2] reported that austenitic stainless steels represent about 60% of the world's total stainless steel production.

Temporary joints of iron metal normally involved with bolt, nut, screw and rivet are susceptible to crevice corrosion because many fine porosity exists between the joints and directly exposed to the atmosphere. To overcome this problem, welding process was preferred. As mentioned by Moniz and Miller [3], gas metal arc welding (GMAW) is a very economical process because it has higher speeds and higher deposition rates than the manual metal arc process, and does not require frequent stops to change electrodes. In addition, this process required very minimum post weld cleaning because the slag is almost absent. These advantages make the process very well adapted to be automated and particularly in robotic welding applications. Nevertheless, it is necessary to observe that welding is one of the sources for the formation of hydrogen embrittlement. Hydrogen can enter the weldment from the parent metal being welded, filler metals and the environment. In the chemical reaction, an atomic hydrogen results from the donation of a proton to a hydrogen ion diffuse through the grains of the alloy matrix to form hydrogen molecules.

According to Izumi et al. [4], the formation of hydrogen molecules create pressure from the cavity that may affect their plastic deformation. Thus, this alloy will experience a decrease in ductility and strength very significantly due to the increased mobility of dislocations and it greatly affects the stress-strain relationship. This argument was supported by Moro et al. [5], where the main effect of hydrogen was to reduce the necking of specimens and the displacement to failure. The aim of our study was to investigate the effects of hydrogen environment to the welding strength on mechanical and microstructural properties of austenitic stainless steel.

14.2   AUSTENITIC STAINLESS STEEL

A material used as a base metal in this experiment is AISI 304 austenitic stainless steel with dimensions of $200 \times 100 \times 5$ mm^3. This plate was analyzed using Spark Optical Emission Spectroscopy test machine (Maxx LMF 14) according to the ASTM A751 standard (American Society for Testing and Materials). The chemical compositions were measured using a spark spectroscopy test machine which containing (wt%) of C: 0.043, Si: 0.378, Mn: 0.96, Cr: 18.21, Ni: 8.13, Mo: 0.006, Cu: 0.054, P: 0.023, S: 0.0042 and balance Fe. Meanwhile, the validation of AISI 304 specimen was performed by comparing the chemical compositions as listed in standard of ASME comprising C: 0.060, Si: 0.460, Mn: 1.23, Cr: 18.31, Ni: 9.05, Mo: 0.300, Cu: 0.25, P: 0.023, S: 0.200 and balance Fe.

14.3 PREPARATION OF WELDING COUPON

The welding coupons were prepared by machining the plate to form a single V-groove of 60° in rolling direction. Surface samples were cleaned with acetone (C_3H_6O) and tacked welded with 2 mm root gap. A copper strip was placed below the joint to avoid accidental welding onto the workbench. The plates were welded in a flat (1G) position by single pass using robotic Gas Metal Arc Welding (GMAW) machine model OTC DR-4000 (Figure 14.1). The filler metal used in this welding procedure was ER308L-Si of 1.0 mm diameter with DCEP polarity. The magnitude of welding heat input is calculated by the following Eq. (14.1):

$$Q = \left(\frac{I \, x \, V}{s} \right) \qquad\qquad (14.1)$$

where Q is welding heat input (kJ/mm), I is current supply (Ampere, A), V is voltage (Volt, V), and s is welding speed (mm/s).

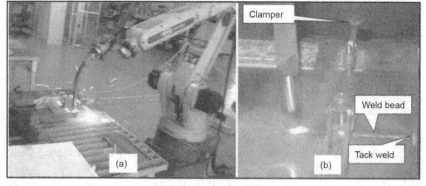

FIGURE 14.1 Welding preparation (a) Robotic GMAW model OTC DR-4000, (b) Welding coupon.

TABLE 14.1   Welding Parameters and Heat Treatment Process of the Specimens

Specimen		Welding Parameters			
Code	Condition	Current (A)	Voltage (V)	Welding speed (mm/s)	Heat input (kJ/mm)
1C2	Heat treated	200	24	3.0	1.600
1C3	Heat treated and H_2 charging				
3C1	Heat treated	*160*	*21*	*2.5*	*1.344*
3C3	Heat treated and H_2 charging				

The magnitude of the welding heat input for both welding coupons that are produced according to the selected parameters as shown in Table 14.1. Then the welding coupons were cut using CNC machine into pieces of standard transverse tensile test shape according to ASTM E8 standard.

14.4   HEAT TREATMENT PROCESS

Five pieces of standard rectangular tensile tests were prepared to perform this experiment. The first specimen was considered as-received (AR) condition specimen of AISI 304 stainless steel without welding. The heat treatments to the specimens were carried out in two stages. In the first cycle, all specimens (4 pieces) were going through solution heat-treated (SHT) at 1035°C for 1 hour and then immediately quenched in water. For the purpose of second cycle heat treatment, the specimens 1C3 and 3C3 which represented of high heat input (1.6 kJ/mm) and low heat input (1.344 kJ/mm), respectively, were tempered in stress relief heat treatment (SRHT) at 300°C for 45 minutes and then allowed to cool in the furnace to room temperature (Table 14.1). Moniz and Miller [3] mentioned that, the SRHT between 649 to 871°C can result in significant distortion and loss of corrosion resistance from sensitization. However, a low-temperature SRHT between 204 to 427°C helps to improve dimensional stability and reduce peak stress, but not reducing corrosion resistance.

The oxide layer formed on the surface of specimen during heat treatment was removed by polishing with fine emery paper and alumina powder. Lastly specimens were etched by Kalling 2 for 20 seconds. The characterization of the microstructures in welding joint was carried out using an Olympus BX60 microscope and the fracture surfaces of the specimens were analyzed by Phenom XL Desktop scanning electron microscopy (SEM).

14.5 EXPERIMENTAL PROCEDURE

For hydrogen environment test, all surface areas of the specimens except the gauge length surface and a shoulder of one end were coated with nail varnish. The coating process was applied to ensure that only the gauge length (50 mm) surface would be exposed to test the solution. The surface of gauge length was cleaned with acetone. The samples were held suspended on glass stand in the glass cell containing 0.3M sulfuric acid (H_2SO_4) solution plus 80 mg per liter of arsenic trioxide (As_2O_3) as shown in Figure 14.2. The end of the specimen which coated with nail varnish was immersed in solution and the other one remaining in covered was connected with alligator clips to the power supply as cathode. In this circuit, platinum gauze electrode 80 mesh was used as anode. A constant current/volt regulated DC power supply (KENWOOD PD 36–10A) was used to control the charging current, which was monitored using digital power

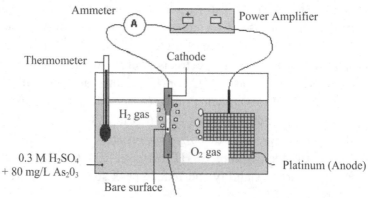

FIGURE 14.2   Experiment setup in glass cell for hydrogen charging test.

quality clamp meter (FLUKE 345). In this study, the specimens (1C3 and 3C3) were charged with constant current density of 110 mA per cm^2 for 12 hours.

According to the study conducted by Costa-Mattos et al. [6] using a slow strain rate of 3.0×10^{-6} s^{-1} its taken about 45 days to break the tensile specimens of AISI 304. In this study, after completing the heat treatment, then followed by hydrogen charging process, the specimen was immediately pulled out until fracture with a slow strain rate of 1.8×10^{-5} s^{-1} using an INSTRON 3382 tensile machine. The percentage decay of ultimate tensile strength and elongation that affected by hydrogen charging can be determined by the following Eq. (14.2):

$$\text{Decay \%} = \left(\frac{\text{Heat treated} - \text{Hydrogen charging}}{\text{Heat treated}} \right) \times 100\% \qquad (14.2)$$

14.6   MICROSTRUCTURE PROPERTIES

Figure 14.3 shows an optical micrograph of AISI 304 stainless steel in the as-received condition (AR) as a base metal in the welding process. Based on the observations of AR specimen, the microstructure of AISI 304 stainless steel has a single phase austenitic structure because no traces of carbides precipitate microstructures found at grain boundaries. This image is quite similar as suggested by William F. Smith [7] and Halil et al. [8].

The actual penetration of weldments which obtained from both cross-sectional of welding coupons are shown in Figure 14.4. The welding coupon welded at 200 A as shown in Figure 14.4(a) has underfill on the face of the bead. However, the weld was acceptable because of the depth of underfill is less than 0.5 mm which smaller than the allowable value of 0.794 mm as mentioned by Sacks [9], referring to API Standard 1104. This means that the underfill defects have a radius but is not a sharp notch which can cause a high stress concentration. Normally, underfill and also undercut may be caused by high welding current, fast welding speed, improper electrode or filler metal manipulation. Welding coupon of 160 A using the slow speed has formed higher weld bead as shown in Figure 14.4(b) compared to welding coupon of 200 A because the liquid of filler metal has more time to solidify.

FIGURE 14.3 Microstructure of AR of AISI 304 after etching using Kalling 2; Mag. 500×.

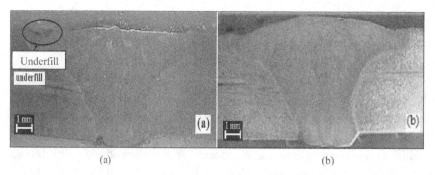

FIGURE 14.4 The cross-sectional of the welding coupon using current supply (a) 200 A and (b) 160 A.

Figure 14.5 shows the magnitude of welding heat input has influenced the formation of the fusion line (FL) thickness. By using high welding heat input has resulted much thicker FL formed and increased the toughness of the weld joint. For example, the welding coupon of 200 A as shown in Figure. 14.5(a) has a thickness of FL about 1,608.5 µm and the thickness of FL in welding coupon of 160 A as shown in Figure 14.5(b) is 1,005.4 µm.

The similar pattern occurs in the formation of the columnar microstructures in the weld metal as shown in Figure 14.6. It is clearly shown that as welding heat input increases the dendrite size, the inter-dendrite spacing in

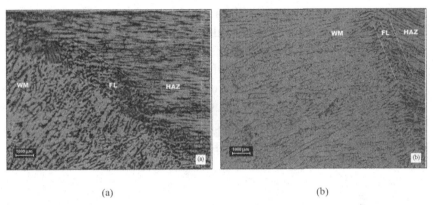

(a)   (b)

FIGURE 14.5   The microstructures of fusion line for welding coupons (a) 200 A; and (b) 160 A; Mag. 100×.

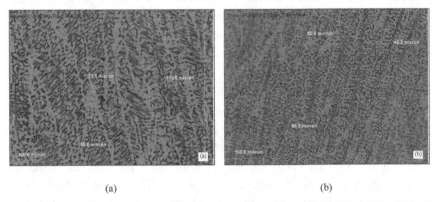

(a)   (b)

FIGURE 14.6   The microstructures of weld metal (a) 200 A and (b) 160 A; Mag. 100×.

the weld metal also increase. For the specimen of welding coupon 200 A as shown in Figure 14.6(a), it takes long time to cool at room temperature. This allows the dendrite formed in the weld metal to become larger. However, for welding coupon 160 A as shown in Figure 14.6(b) required a short time to equaling the surrounding temperature caused dendrite size becomes smaller and the inter-dendrite spacing was getting close. The average values of columnar crystals for welding coupons of 200 A and 160 A are 91.96 and 49.23 μm, respectively. Similar results were produced by Subodh and Shashi [10] in his analysis of the TIG welding on AISI 304 type.

The precipitates that exist in the austenite matrix of weld metal can be classified into two types, namely lathy ferrite in the columnar grain and vermicular (skeletal) ferrite. The existence of this precipitate shows that the alloy contains a lot of ferrite. Lathy ferrite microstructure can emerge due to greater ferrite contents or a characteristic cooling time after the welding procedure. Based on an investigation conducted by Sathiya et al. [11], a fully dendrite microstructure consists of a darker austenite phase is primary which called dendrite and lighter one is secondary inter-dendrite phase.

14.7   TENSILE TEST PROPERTIES

AISI 304 stainless steel is the most popular due its high ductility and weld-ability. The values of the ultimate tensile strength (UTS) and percentage of elongation (% EL) for AR specimen were 732.89 MPa and 61.16%, respectively. As shown in the Figure 14.7(a), AR specimen has high duc-tility features because the existence of necking formed in the fractured area. However, after welding at 200A (1C2) and 160A (3C1), respec-tively, and followed by heat treatment process, their percentage elonga-tion were reduced (Table 14.2), were approximately of 40.4% and 19.8%, respectively. All tensile test specimens were fractured in the weld joints a shown in Figures 14.7(b) and (c). This indicates that the base metal is stronger than weld metal. Such conditions may occur due to inadequate weld penetration at the root of the weld. After hydrogen charging test, the specimens that were produced using high heat input (1C3) and low heat

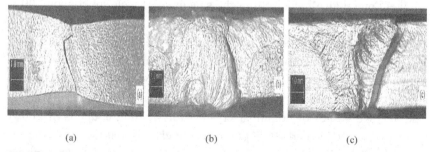

(a)   (b)   (c)

FIGURE 14. 7   The fractured region of tensile test specimens, (a) AR, (b) 1C3, and (c) 3C3.

TABLE 14.2 Tensile Properties of AISI 304 Stainless Steel Affected by Hydrogen Charging

Specimen	AR	1C2	1C3	3C1	3C3
UTS (MPa)	732.89	645.4	418.7	407.9	322.3
% Elongation (G.L. = 50 mm)	61.16	40.4	20.4	19.8	9.4

input (3C3) have experienced some brittleness because of a broken part looks straight compared to the AR specimens.

Figure 14.8(a) shows that the strength of welding coupon produced by high heat input welding is stronger than coupon welded at low heat input. However, after going through the hydrogen charging process with current density of 110 mA/cm² for 12 hours, the value of UTS and % EL for both welding coupons were decreased. By applying Eq. (14.2) into Table 14.2, the percentage of decay of UTS for welding coupon 200 A about 31.12% and 20.99% for welding coupon 160 A. The similar trend occurring in elongation, where the decay in elongation for welding coupon 200 A and 160A are 49.50% and 52.53%, respectively as shown in Figure 14.8(b).

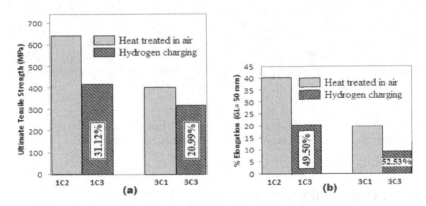

FIGURE 14.8 Decreasing of tensile properties of AISI 304 affected by hydrogen charging: (a) Ultimate tensile strength, and (b) Percentage of elongation. Note that the percentage within parentheses shows the decay of UTS and ductility.

14.8   FRACTURE SURFACES

All fractured surfaces of the standard rectangular tensile test specimens were examined using Phenom XL Desktop SEM. The fractured surface of AR specimen (Figure 14.9(a)) is free from the effects of the globular oxides which make the AR specimen has mechanical properties (strength, toughness and ductility) are better than welded specimens. There are more globular oxides formed in coupon welded at 200 A (heat input of 1.64 kJ/mm) supplied by hydrogen charging compared to coupon welded at 160 A (heat input of 1.45 kJ/mm) as shown in Figures 14.9(b) and 14.9(c), respectively. The formation of globular oxides [12] is shown to have some influence on the mechanical behavior of weld joints. Globular oxide acts as a starting point of hydrogen gas pressure in welding joints, fine cracks produced at the oxide can weaken the mechanical properties eventually make the specimens brittle and prone to fracture.

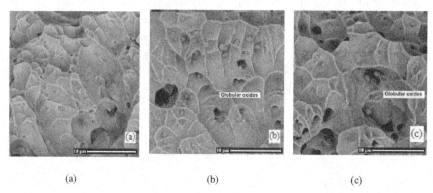

(a)   (b)   (c)

FIGURE 14.9   Image of fractured surfaces of the standard rectangular tensile tests specimen viewed under SEM. (a) AR, (b) 1C3, and (c) 3C1.

14.9   CONCLUSION

The increased heat input during welding may increase the strength of weldment due to increase in dendrite size and inter-dendrite spacing. However, after going through hydrogen charging process, the strength was decreased. The specimens that were charged with hydrogen revealed the characteristic of lower ductility than heat treated specimens. All the welded specimens fractured at the weld joint, they were found to have globular oxides but the AR condition specimen was free from them. The presence of globular oxides on the fracture surface of weldment for the specimen be charged with hydrogen showed that the hydrogen embrittlement phenomenon has a significant influence on the mechanical properties of austenitic stainless steels.

KEYWORDS

- **austenitic stainless steel**
- **fracture surfaces**
- **heat treatment**
- **microstructure**
- **tensile test**
- **welding coupon**

REFERENCES

1. Henkel, D., & Pense, A. W. (2001). *Structure and Properties of Engineering Materials*, 5th ed., McGraw Hill.
2. George, E. T. (2006). *Steel Heat Treatment Handbook*, 2nd ed.; Taylor and Francis Group.
3. Moniz, B. J., & Miller, R. T. (2004). *Welding Skills*, 3rd ed; An ATP Publication.
4. Izumi, T., Itoh, G., & Itoh, N. (2004). Hydrogen Permeation Behavior in Aluminium Alloys. *Materials Science Forum. 28*, 751–757.
5. Moro, I., Briottet, L., Lemoine, P., Andrieu, E., Blanc, C., & Odemer, G. (2010). Hydrogen Embrittlement Susceptibility of a High Strength Steel X80, *Materials Science and Engineering A. 527*, 7252–7260.
6. Costa-Mattos, H. S., Bastos, I. N., & Gomes, J. A. C. P. (2008). A Simple Model for Slow Strain Rate and Constant Load Corrosion Tests of Austenitic Stainless Steel in Acid Aqueous Solution Containing Sodium Chloride, *Journal of Corrosion Science, 50*, 2858–2866.

7. William F. S., (2004). *Foundations of Materials Science and Engineering*, 3rd ed., McGraw Hill.

8. Halil, I. K., & Ramazan, S., (2013). Study on Microstructure, Tensile Test and Hardness 304 Stainless Steel Jointed by TIG Welding, *International Journal of Science and Technology*, 2(2), 163–168.

9. Sacks, J. R., & Bohnart, E. R., (2005). *Welding Principles and Practices*, 3rd ed., McGraw Hill.

10. Subodh, K., & Shahi, A. S. (2011). Effect of Heat Input on the Microstructure And Mechanical Properties of Gas Tungsten Arc Welded AISI 304 Stainless Steel Joints, *Materials and Design, 32*, 3617–3623.

11. Sathiya, P., Aravindan, S., Ajith, P. M., Arivazhagan, B., & Noorul Haq, A. (2010). Microstructural Characteristic on Bead on Plate Welding of AISI 904L Super Austenitic Stainless Steel Using Gas Metal Arc Welding Process, *International Journal of Engineering, Science and Technology*, 2(6), 189–199.

12. Galvis, E. A. R., & Hormaza, W. (2011). Characterization of Failure Modes for Different Welding Processes of AISI 304 Stainless Steels, *Engineering Failure Analysis, 2*, 1791–1799.

CHAPTER 15

COMPRESSION STRENGTH OF OPEN-CELL ALUMINUM FOAM PRODUCED BY GREEN SAND CASTING USING THE ENERGY CONSERVATION LOST-FOAM TECHNIQUE

MOHD KHAIRI TAIB, RAZMI NOH MOHD RAZALI, BULAN ABDULLAH, and MUHAMMAD HUSSAIN ISMAIL

Faculty of Mechanical Engineering, Universiti Teknologi MARA, Shah Alam, Selangor, Malaysia

CONTENTS

OVERVIEW

The purposes of this study were to determine the mechanical behavior of aluminum foam by compression test and the porosity percentage of the three different types of aluminum foam produced by energy conservation lost foam method. Commercial polystyrene usually used for packaging was chosen as a template and the desired templates were produced and filled with sand, which was then embedded in green sand casting (recycle material). The mold subsequently poured with a molten aluminum based on reusing and the samples was then undergoing blasting process to remove the sand. It was followed by a machining process to remove surface imperfections during casting. The mechanical behavior was investigated under uniaxial compression with solid aluminum and three different arrangements of pore structure with equal size samples of Al Foam 4, Al Foam 9 and Al Foam 14. As a result, the aluminum foams' mechanical behavior differs with different pore size and porosity. The higher its density will resultant higher ability to absorb energy.

15.1 INTRODUCTION

Metallic foams are new kinds of material that demonstrate attractive characteristics when compared to their solid material. Earlier, pores defect is considered unacceptable for engineering requirements. However, in recent years, a great significance has been placed on a new class of engineering material, known as "Porous metals" [1]. This new class of engineering material has distinctive mechanical and physical properties that offer low densities and novel physical, mechanical, thermal, electrical and acoustic properties [1, 2]. Various methods have been established in producing aluminum foam. Most of them used casting approach, powder metallurgy and foaming from liquid as their main processes. The only difference between them is their method on how to produce the foams or the pore structure.

Currently, there is certainly a growing interest in making use a aluminum foams for their advantages, such as comparatively high stiffness regardless of very low density, high deterioration level of resistance, outstanding noise absorption as well as vibration reductions and also easy recycling where possible [3]. Aluminum foams contain 20–95% pores show excellent performance with regards to energy absorption with respect to polymeric foams, regardless of their greater weight per unit volume [4]. As an example, aluminum foams possess a range of allowable temperatures, which is five times larger than that of polymer foams, which have a maximum temperature limit of approximately around 100°C [3, 4]. Due to demand from industries, the studies regarding on finding ways to improve the existing aluminum foam shows a significant progress day to day.

15.2 LOST FOAM METHOD

Lost foam uses an expendable pattern made of molded polymer usually commercial polystyrene. The pattern is coated with a refractory material and placed inside a steel flask, where it is surrounded with loose, dry sand. After the sand is compacted by vibration, liquid metal is poured directly into the pattern, which decomposes ahead of the advancing liquid metal as gas and liquid products from the receding foam diffuse through the porous coating and into the sand. The liquid metal eventually replaces all the volume occupied by the foam pattern before it solidifies [5].

There are various methods had been developed to produce aluminum foam depending on its applications which ranging from conventional casting to powder metallurgy routes. Several methods exist for the manufacture of aluminum foam, namely melts-gas injection, melt-foaming, powder metallurgy, investment casting and melt infiltration [6–12]. However, by lost foam method several benefits like getting rid of machining engagements, making complicated casting without having cores and decreasing environmental loads could be offered through this procedure [4]. Furthermore, lost foam method was reported to save up to 30% of energy compared to traditional sand casting. In a short period of time, these costs, saving method increase more than 100% using aluminum since 1986 [13].

15.3 OPEN-CELL METAL FOAM

In general, there are two types of metal foam. The two types consist of open-cell and closed-cell metal foam. In this study, open-cell aluminum foams are being fabricated. Open-cell metal foam has an interconnected system of solid struts, which permits fluid to pass from end to end. Therefore, applications such as heat exchanger filter and medium is suitable whereas it allows the passage of fluid and also gasses [1]. The particular mechanical properties of metallic foam are firmly influenced from the porous structure and the degree of interconnectivity of each and every pore cell [1]. Lower density will only increase its ability to absorb energy [14]. Furthermore, it also reflects on the lost foam method which clearly low in cost and environmental friendly, the mechanical property depends on the design of the metal foam prior from the design of polymeric foams template (replicated the structure of molten metal). The energy absorption can be determined under the stress-strain plateau [1, 2].

15.4   CHEMICAL COMPOSITION DETERMINATION

The aluminum ingot used for the casting of the aluminum foam by Green Sand Casting Using Lost-Foam Technique was determined its chemical composition using spectrometer test. The aluminum ingot was tested three times on the same surface and the average chemical composition was taken.

15.5   MECHANICAL AND PHYSICAL TESTING

Three different designs of aluminum foam were fabricated through lost foam method. The arrangements of pore structure of aluminum parameters are aluminum foam with 4 holes (Al-form4), aluminum foam with 9 holes (Al-form9) and aluminum foam with 16 holes (Al-form16). The holes are in same in terms of size, which is 2 cm in diameter. Green sand was selected as the mold which contained cope and drag (represent upper and lower part of the mold). Then, the polystyrene with desire shape was buried into the mold before conventional casting took place.

Compression test was performed to determine the stress-strain curve, and the compressive strength of aluminum foam. Basically, the process involved the placement of samples in between two compression plates and load was applied on top of it (Uniaxial). The samples were machined using lathe machine to achieve final dimension of 43 ± 1 mm and 44 ± 1 mm in diameter and height, respectively. The samples need to be flat prior compressed in order to avoid error or misalignment of the specimen during compression.

The compression machine used to determine the mechanical behavior was Universal Testing Machine Digital Servo Control. The aluminum foam was placed between the jig and the speed rate was applied using Instron Bluenil software. The speed rate of the compression test for aluminum foam was compressed with the rate of 1.2 mm/s. Every type of aluminum foam underwent the same parameter.

The energy absorption was determined using MATLAB software. In addition, the area under the graph was calculated using the trapezoidal rule.

Since the volumes of the samples are known by determining using volume equation after measuring the mass, width and height of the samples. The dimensions and volumes were measured and determine using a Vernier caliper while each of the sample's weight was weigh using digital weigh machine. The density of the product and percentage of porosity can be calculated using Eqs. (15.1) and (15.2):

$$\text{Density, } \rho = \frac{\text{Mass, m}}{\text{Volume, V}} \qquad (15.1)$$

$$\text{Percentage porosity, } \% = 1 - \frac{\text{New density, } \rho n}{\text{Al density, } \rho} \times 100 \qquad (15.2)$$

15.6   RESULTS AND DISCUSSION

15.6.1   CHEMICAL COMPOSITION PROPERTIES

The chemical composition of aluminum dense was determined using spectrometer test. The result is shown as in Table 15.1. It clearly shows that the material is not purely aluminum since the percentage of aluminum composition is only 76.9%. Using a recycle material was believed to be the reason of the small amount of aluminum exists compared to pure aluminum. Another component such zinc and silicon exhibit high existence which is

TABLE 15.1   Chemical Composition of Dense Aluminum

Material	Si	Fe	Cu	Zn	Sn	Zr	Ca	Al
Percentage of Composition (%)	7.58	2.16	2.00	10.74	0.021	0.0043	0.0061	76.90

7.58% and 10.74%, respectively. This may due to the use of same crucible in every casting. It is observed that the same crucible was used repeatedly for other casting which using different material.

15.6.2   COMPRESSION BEHAVIOR

The compression curve can be divided into 3 stages as shown in Figure 15.1. The first stage is a linear elastic deformation stage; second stage is a plateau deformation stage and thirdly densification stage [1]. Al-foam4 exhibit the higher compressive stress compared to aluminum foam. It clearly shows that number of porosity and cell thickness significantly influences the strength of the aluminum. Al-foam16 with the percentage of porosity, 48.52% and cell thickness, 3 mm has the lowest strength compared to its counterparts.

From Figure 15.2, the aluminum foam with four holes can be seen deformed for every 10% of strain. It can be considered as fragile and started to rupture at some point of as early of 30% of strain. This is believed due to the big pore size of interconnected cellular of aluminum foam. Therefore, big pore size of aluminum foam can withstand compressive pressure for a short period of time. Meanwhile, observation on Al-foam9 and Al-foam14 also tend to be fragile and started to rupture at some point, respectively at

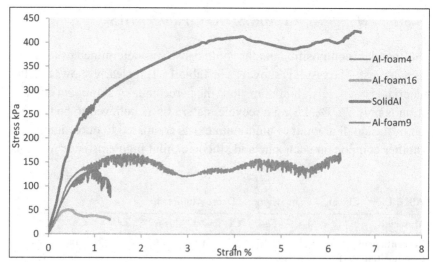

FIGURE 15.1   Compressive stress-strain curve of solid aluminum and aluminum foam.

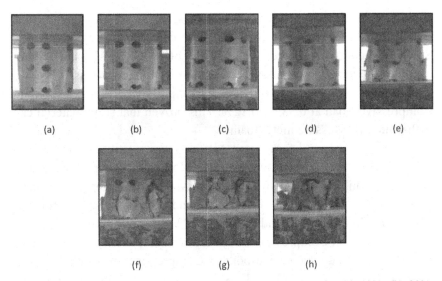

(a)   (b)   (c)   (d)   (e)

(f)   (g)   (h)

FIGURE 15.2 Deformation of Al-foam4 at every 10% of strain: (a) 10%, (b) 20%, (c) 30%, (d) 40%, (e) 50%, (f) 60%, (g) 70%, (h) 80%.

0.889% strain and 0.389% strain. Conversely, both of it failed on the early stage of the compression test.

Other than that, Table 15.2 shows solid aluminum possesses the highest maximum compressive load of 625.00 ± 23.00 kN, Al-foam16 has the lowest maximum compressive load of 76.612 ± 2.00 kN. Additionally, Solid

TABLE 15.2 Compression Properties of Solid Aluminum and Aluminum Foams

Type	Maximum Compressive Load (KN)	Compressive Stress at Maximum Compressive Load (KPa)	Compressive Strain at Maximum Compressive Load (%)	Modulus (MPa)
SolidAl	625.00 ± 23.00	434.00 ± 18.00	6.66 ± 0.33	405.90 ± 9.00
Al-foam4	247.54 ± 26.00	170.455 ± 18.00	2.814 ± 5.00	346.68 ± 12.00
Al-foam9	209.23 ± 23.00	144.078 ± 23.00	0.889 ± 23.00	203.36 ± 23.00
Al-foam16	76.612 ± 2.00	53.571 ± 5.00	0.389 ± 0.12	184.74 ± 68.00

aluminum demonstrates the highest compressive stress at maximum compressive load of 434.00 ± 18.00 kPa which corresponds to Compressive Strain at Maximum Compressive Load at 6.66 ± 0.33%. Al-foam16 demonstrates the lowest compressive stress at maximum compressive load of 53.571±5.00 kPa which corresponds to Compressive Strain at Maximum Compressive Load at 0.389 ± 0.12%. This proven that solid material can withstand high pressure metal foam.

Aluminum foam Al-form4 and Al-form9 has relatively small difference in terms of Compressive Stress at Maximum Compressive Load. Al-form4 exhibit Compressive Stress at Maximum Compressive Load of 170.455 ± 18.00 kPa which correspond to Compressive Strain at Maximum Compressive Load of 0.889 ± 23.00%. Aluminum foam type Al-foam 9 exhibit Compressive Stress at Maximum Compressive Load of 53.571 ± 5.00 kPa which correspond to Compressive Strain at Maximum Compressive Load of 0.389 ± 0.12%. Therefore, the solid aluminum owns the highest modulus of 405.90 ± 9.00 MPa and the lowest is Al-form16 where the modulus are 184.74 ± 68.00 MPa. It is believed that the more holes in the material, the fragile it will be.

It clearly shows that the relationship between elastic modulus and the foam is inversely proportional. The elastic modulus decreased with the increasing of the foam. SolidAl exhibited the highest strength of modulus elasticity compared to aluminum foam. The modulus of Al-foam4 is much higher compared to Al-foam 16. This was caused by the cell wall thickness which is the thickest the cell wall is, the higher the modulus. In other words, the higher thickness of cell wall would result in foams with greater strength and modulus elastic

15.6.3 ENERGY ABSORPTION

The energy absorption was calculated in a range of 0% strain and 6.259% and the results are shown in Figure 15.3. It clearly shows that from the stress-strain curve, the total energy absorption is increased with increasing number of strains. This is because the energy absorption was measured by calculating the area under plateau stress graph [13]. However, this type of trends cannot be seen at Al-foam 16 as it exhibits a small number of energy absorption. This is due to a small length of plateau stress. Al-foam16 faced an early failure at strain 1.22%, while Al-foam4 still exhibit densification at strain 6.259%.

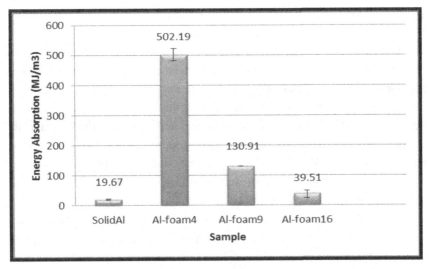

FIGURE 15.3   Average of energy absorption of aluminum foam for all densities.

This was believed to be due to internal crack of the foam since their cell wall is smaller compared to Al-foam4. This small thickness of cell wall could not able to withstand the load applied to it and caused the structure to collapse and failure at a low compressive strain [11]. Another reason, maybe because of the cell structure is damaged during filling in the sand in the pattern. Since the wall thickness of pattern is small, the tendency to damage during compaction of sands is high due to its polymer behavior which is very soft. This caused no strut wall in some pore structure.

This resulted in the high value of the area under graph which represents the energy absorption capability. It can be said that, Al-foam4 can support higher stress before fracture. It was proved that, the energy absorption, increased as the porosity increased. However, some cases excluded from this theory, such Al-foam16. Al-foam16 exhibited the small number of energy absorption even though it has a higher percentage of porosity. This is because of Al-foam16 has a thinner cell wall. These thin walls could not support the load and failed at low stress and strain. Since the area under the stress-strain curve represents the amount of energy absorbed.

Solid aluminum possesses the lowest energy absorption of 19.67 MJ/m³. This is due to the energy absorption of a cellular solid is characterized by the hatched area up to a stress its termed the peak stress (σ p), which indicates the transition to the densification regime at about 1.4% strain [1, 2].

15.7   CONCLUSION

The lost foam technique is a low cost method and conserves a high amount of energy when compared to the conventional sand casting process. It has completely improved industry sectors in energy consumption and so on.

From the experiment, it has been seen that aluminum foam with low density possess a higher amount of energy absorption while low density aluminum foam tends to possess lower energy absorption. Therefore, amount of holes in aluminum foam tend to design can be predicted whether high or low energy absorption depending on the amount of holes. The higher amount of holes in aluminum foam, the lower its density.

This study focused on the effect of porosity percentage and mechanical behavior of aluminum foam produced by recycle, green sand casting process using energy conservation of lost-foam technique to sustain energy. The following conclusions can be made from the outcome of this study:

1. The melt aluminum replicates the shape of the polymer pattern to produce aluminum foam.
2. The thickness of the cell wall and porosity influence the strength of aluminum foam.
3. Al-foam4 exhibited the highest energy absorption as it has a long plateau stress. While, Al-foam16 shows the lowest energy absorption.

ACKNOWLEDGMENTS

The authors expressed their gratitude to Faculty of Mechanical Engineering and Research Management Institute (RMI) Universiti Teknologi MARA, Shah Alam, Malaysia for financial support. The authors also would like to thank all the undergraduate students, faculty members and all those involved in this project.

KEYWORDS

- aluminum foam
- energy absorption
- lost foam method

- **open-cell metal foam**
- **recycle**
- **sand casting**

REFERENCES

1. Ashby, F. M., Evans, A., Gibson, L. J., Hutchinson, J. W., & Wadley, H. N. G. (2000). *Metal Foam: A Design Guide;* Butterworth-Heinemann.
2. Degischer, H.-P., & Kriszt, B. (2002). *Handbook of Cellular Metals: Production, Processing, Applications,* Wiley-VCH Verlag GmbH & Co. KGaA.
3. Kim, K., & Lee, K. (2005). Effect of Process Parameters on Porosity in Aluminum Lost Foam Process. *J. Mater. Sci. Technol. 21*(5).
4. Caulk, D. A. (2006). A Foam Melting Model for Lost Foam Casting of Aluminum. *International Journal of Heat and Mass Transfer, 49*, 2124–2136.
5. Simančík, F., Jerz, J., Kováčik, J., & Minár, P. (2003). Aluminum Foam – A New Light Weight Structural Material. Institute of Materials and Mechanics, SAS, Raèianska.
6. Dawood; Shaik, A. K., & Nazirudeen, S. S. M. (2010). A Development of Technology for Making Porous Metal Foams Castings. *Jordan Journal of Mechanical and Industrial Engineering, 4*(2), 292–299.
7. Surace, R., Filippis, L. A. C. D., Ludovico, A. D., & Boghetich, G. (2009). Influence of Processing Parameters on Aluminum Foam Produced by Space Holder Technique. *Materials and Design. 30*, 1878–1885.
8. Shi, Zhao, W. (2006). Effect of Y2O3 on the Mechanical Properties of Open Cell Aluminum Foams. *Materials Letters.* pp. 1665–1668.
9. Peroni, L., Avalle, M., & Peroni, M. (2008). The Mechanical Behavior of Aluminum Foam Structures in Different Loading Conditions. *Int. J. Impact Eng. 35*(7), 644–658.
10. Michailidis, N., & Stergioudi, F. (2011). Establishment of Process Parameters for Producing Al-Foam by Dissolution and Powder Sintering Method. *Mater. Des. 32*(3), 1559–1564.
11. Bafti, H., & Habibolahzadeh, A. (2010). Production of Aluminum Foam by Spherical Carbamide Space Holder Technique: Processing Parameters. *Mater. Des. 31*(9), 4122–4129.
12. U.S. Department of Energy. (2013). *MiChigan. Energy Efficiency & Renewable Energy,* 1–9.
13. Grilec, K., Marić, G., & Katica, M. (2012). Aluminum Foams in the Design of Transport Means. *Promet – Traffic Transportation, 4*(24), 295–304.

INDEX